T0329684

VITAL DECOMPOSITION

VITAL DECOMP-

SOIL PRACTITIONERS + LIFE POLITICS

OSITION

Kristina M. Lyons

Duke University Press   Durham and London   2020

Printed in the United States of America on acid-free paper ∞
Designed by Drew Sisk.
Typeset in Mimion Pro, Avenir, and Canela Text
by Copperline Book Services.

Library of Congress Cataloging-in-Publication Data
Names: Lyons, Kristina M. (Kristina Marie), [date] author.
Title: Vital decomposition : soil practitioners and life politics /
Kristina M. Lyons.
Description: Durham : Duke University Press, 2020. | Includes
bibliographical references and index.
Identifiers: LCCN 2019032478 (print) | LCCN 2019032479 (ebook)
ISBN 9781478007692 (hardcover)
ISBN 9781478008163 (paperback)
ISBN 9781478009207 (ebook)
Subjects: LCSH: Soil degradation—Colombia—Putumayo (Department) |
Environmental degradation—Colombia—Putumayo (Department) | Nature—
Effect of human beings on—Colombia—Putumayo (Department) | Putumayo
(Colombia : Department)—Environmental conditions.
Classification: LCC S625.C7 L96 2020 (print) | LCC S625.C7 (ebook) |
DDC 631.4/520986163—dc23
LC record available at https://lccn.loc.gov/2019032478
LC ebook record available at https://lccn.loc.gov/2019032479

Cover art and frontispiece: Soil © Image Source.
Courtesy Getty Images/Image Source.

Para mi gran familia andinoamazónica,
seguimos soñando despiertxs, caminando juntxs y tejiendo vida.
For teaching me how to live as more than one and less than two.

CONTENTS

ACKNOWLEDGMENTS

This book emerged from the generosity of many conversations, practices of cultivation, hikes, and life. I would first like to thank my extended network of family, friends, and compañerxs in the Andean-Amazonian foothills and plains—in particular, Heraldo Vallejo, Nelso Enriquez, María Elva Montenegro, Carlos Becerra, Jorge Luis Guzmán, Elizabeth Guzmán, Edgar Torres, Constanza Carvajal, Rocio Ortiz, Casa Amazonia, Selva Morena, and the members of Fundación Kindicocha, Mesa Regional de Organizaciones Sociales Indígenas, Campesinas y Afros del Putumayo, Baja Bota Caucana y Cofanía Jardines de Sucumbíos, Red de Guardianes de Semillas de Vida del Sur, and Fundación ItarKa. Also essential were colleagues at the agricultural microbiology laboratory of the National University's Institute of Biotechnology in Bogotá, especially Javier Vanegas, Daniel Uribe, Giovanna Landazábal, Alexandra Sanchéz, Anyela Rodriguez, and Nathalia Floréz; the Geographic Institute Agustín Codazzi and the National Soil Science Laboratory, especially Abdón Cortés, Julián Serna, Jorge Sanchéz, Ricardo Siachoque, Óscar Avecedo, and Diana Vera; the Corporación para el Desarrollo Sostenible del Sur de la Amazonia, especially conversations with Iván Melo, Mauricio Valencia, and Guillermo Martínez; the Colombian Society for Soil Science, especially Ibonne Valenzuela and Samuel Caicedo; and Armando Castilla at Fedearroz, Carolina Olivera at the FAO, and members of the Colombian Association of Agrologists. I am also indebted to colleagues in several nonprofit organizations, especially Asociación MINGA, Witness for Peace, and Acción Ecológica, for inviting me to accompany their long-term work to support the human rights and environmental struggles of communities in Putumayo and Sucumbíos, Ecuador. I would also like to give special thanks to María Clemencia Ramírez for her kindness and for generously sharing her extensive academic and life experience in Putumayo; Efrén Piña Rivera for our many conversations about alternatives to development in southern Colombia; and Tomás Leon for sharing his love of soils and agroecology, and his expertise on the genealogy of Colombian soil science.

I received instrumental feedback on this manuscript from a variety of people. I am sure that I am forgetting some essential voices, but let me mention Jake Culbertson, Tanya Richardson, Kregg Hetherington, Fatih Tatari, and María Puig de la Bellacasa. Two reviewers at Duke University Press read through the entire manuscript, and their insights greatly helped to organize and deepen the material. Early drafts received the invaluable comments of my dissertation committee and advisers at the University of California, Davis. I thank Suzana Sawyer for her graceful articulations and scholarly insights; Joe Dumit for sharing his experimental approach to life and for being such an amazing listener and versatile scholar; Alan Klima for his deep commitment to writing and for his reminders to be watchful of translation in my attempts to stay with the poetics; Kate Scow for the endless nights of soil whispering and for being open and generous with interdisciplinary exchange and collaboration. Marisol de la Cadena has been an endless source of inspiration and support. Thank you for sharing how you think and that which impassions you, for your unflinching vote of confidence, and your relentless impulses and tender and caring bites. I also thank my colleagues and students at the University of California, Santa Cruz, for creating such a vibrant intellectual environment in which to think and write, especially my colleagues affiliated with the Science and Justice Research Center: Donna Haraway, Karen Barad, Jenny Reardon, Lindsey Dillon, Kate Darling, and Colleen Massengale. Andrew Mathews was key in supporting my UC President's Postdoctoral Fellowship application and award. I was so lucky to work with Kali Rubaii, Krisha Hernández, Darcey Evans, Zahirah Suhaimi-Broder, Stephanie McCallum, Daniel Schniedewind, and Rachel Cypher in their capacity as student-colleagues.

Versions of arguments and sections of this book were shared with and benefited from a number of audiences. These include audiences at the Departments of Anthropology at the University of Chicago; Tufts University; and University of California, Irvine; the Our Own Devices Workshop organized by the Graduate Institute for Design, Ethnography and Social Thought and the Parsons School of Design at the New School; the Resistance Is Fertile: On

Being Sons and Daughters of Soil Workshop at the University of Cape Town, South Africa; the Rediscovering Soils: Knowledge and Care in the Worlds of Soil Workshop organized by the University of Sheffield, England; the Environmental Studies Colloquium, Latin American and Latino/a Studies Symposium, Science and Justice Research Center, and Ethnographic Engagements Workshop in the Anthropology Department at the University of California, Santa Cruz; the "Culture, Power, and Social Change" lecture series at the University of California, Los Angeles; the Science Studies Program Colloquium Series at the University of California, San Diego; the Technoscience Salon and Women and Gender Studies Institute at York University; the launch of the National Policy for Sustainable Soil Management organized by the Ministry of the Environment and Sustainable Development and the Food and Agriculture Organization of the United Nations in Bogotá, Colombia; the seminar space of the former Center for Study in Political Ecology in Bogotá, Colombia; the weekly seminar of the Agricultural Microbiology Laboratory of the National University's Institute of Biotechnology in Bogotá, Colombia; and the Ethnographic Museum of the Cultural Deputy of the Bank of the Republic in Leticia, Amazonas, Colombia.

Life is relation, and I am blessed to have an amazing network of cultivated kin. I am infinitely grateful for the love and support of my dear sister Oneida Giraldo and the entire Giraldo-Camargo family; the deep friendship of Astrid Flórez, Raquel Díaz, Felicity Aulino, Karina Hof, Rachel Cypher, Marcela Cely Santos, Catalina Giraldo, and Juan Diego Prieto; the collaborative impulses of my inspiring feminist colleagues Tania Pérez-Bustos, Lina Pinto-García, Alejandra Osejo, Carolina Botero, and Juana Dávila; the kindred spirit of Diana Bocarejo; and the hermandad and intellectual companionship of Iván Vargas Roncancio. To Tom Chauvin, I thank you for your unconditional loyalty and for introducing me to anthropology. My heartfelt thanks to the Ganley-Roper family for welcoming me into your fold, and especially to Barbara for encouraging me to embrace ethnographic poetry. I will be forever grateful to Sylvia Sensiper for her generous soul, and for embracing me in a dark hour. Last but not least, I am thankful for my dear father, Everett Lyons, and my graceful, stubborn, protective angel of a grandmother, Evelyn Pilbin-Lyons, whose support in this world and beyond has sustained me each and every day. To my mother, I am sorry that you were not here to see this publication come to fruition, but I trust that you are making a heart gesture in the air with your hands.

Research for this book was made possible by funding from the Wenner-Gren Foundation, the Social Science Research Council, the University of California Pacific Rim Research Program, and the University of California President's Postdoctoral Fellowship Program. Writing was facilitated by the Andrew W. Mellon Foundation and University of California, Davis, Sawyer Seminar: Indigenous Cosmopolitics: Dialogues about the Reconstitution of Worlds, and a generous Hunt Postdoctoral Fellowship from the Wenner-Gren Foundation.

LIFE IN THE MIDST OF POISON

During the years that I engaged in my long-term fieldwork in southern Colombia, the military obliged people to periodically stop at roadside checkpoints. Depending on the public order scenarios, these stops could occur every forty-five minutes. At best, they happened only three times during the overnight transit from the Amazonian state of Putumayo to the Andean capital city of Bogotá. Generally a thirteen- to sixteen-hour bus ride, the trip depends on seasonal weather cycles and uneven road conditions over what are categorized as geologically unstable fault lines known to produce rock and mud slides during heavy tropical rains. The routine at the checkpoints was always the same.

> *National ID card out.*
> *Boots off.*
> *Backpack open.*
> *(Labrador sniffing)*
> *Don't smile at him (the dog). Don't even look at him.*
> *Where to? Where from?*

Often the soldiers handed out slips of paper that read: "Guerrillas in the 32nd and 48th Fronts of the FARC, demobilize now! Your family misses you. Come back to them. Come back to us. Live with dignity!"

Whenever I return to the memory of these scenes, I hear the sound of soldiers cutting open people's packages. Amid the crates of hens, plastic bags, and small suitcases stored beneath the bus, these packages contained homemade cheeses or other foods and products from farms that had been prepared as gifts for family members living elsewhere in the country. There were always murmurs of protest emanating from the passengers, but mostly quiet resignation. Arms folded across chests as people watched the soldiers slice open their blocks of cheese and rustle through duffle bags in search of presumed hidden bags of cocaine. There was a repetitive and mundane violence unleashed in these acts, the sound of knives splitting open packing tape and cutting across hand-tied string.

When I first traveled to the state of Putumayo in 2007, I participated in a policy watch and human rights delegation with the US-funded NGO Witness for Peace, or, as it is known in Colombia, Acción Permanente por la Paz. The delegation was organized to track the impacts of US antidrug policy and its entanglements with what was, at the time, Colombia's ongoing fifty-year-plus war between the national government and the longest-standing leftist guerrilla organization in the Western Hemisphere, the Revolutionary Armed Forces of Colombia–People's Army (FARC-EP). Often referred to as the gateway to Colombia's Amazonia, Putumayo shares borders with Ecuador and Peru and transitions from central Andean foothills to the extensive Amazonian plains that make up 85 percent of its territory. In 2000, the year the bilateral US-Colombia counternarcotics policy Plan Colombia commenced, Putumayo produced approximately 40 percent of the country's illicit coca cultivations (UNODC 2005). The region was quickly converted into the epicenter of bilateral militarized eradication and state and USAID illicit crop substitution programs.

Between the 2000 and 2012 fiscal years, the US Congress appropriated more than $8 billion to carry out Plan Colombia. As much as 80 percent of this assistance was invested in providing weapons, equipment, infrastructure, personnel, and training for Colombian military and police, including contracts with US-based multinationals that form part of the military-industrial complex, such as Monsanto, Sikorsky Aircraft, and DynCorp International (Beittel 2012). Since the late 1970s, illicit crop eradication strategies in Colombia have utilized chemical warfare tactics, including the application of paraquat, Garlon 4, Imazapyr, and Tebuthiuron.[1] By 2000, the policy relied on a controversial aerial fumigation program involving crop-duster planes that sprayed a concentrated formula of Monsanto's herbicide glyphosate over suspected illicit marijuana, coca, and opium poppy plants.[2] Given the volatile nature of aerial dispersal as a chemical application method, pastures, forests, soils, livestock, subsistence foods, watersheds, and human bodies were regularly misted with glyphosate. More than 1.8 million hectares of coca were aerially fumigated after 1994, including 282,075 hectares in Putumayo since 1997.[3] Despite the generalized failure to reduce the overall quantity of illicit coca, this policy lasted until October 1, 2015, several months after the Colombian government passed a national resolution to officially suspend the aerial spraying of glyphosate.[4] The resolution (006) came in the wake of a report published by the World Health Organization's cancer research arm declaring the world's most widely used herbicide a *probable* carcinogen in humans. Regardless of this suspension, Colombia's Narcotics Council approved the continued manual application of the herbicide to eradicate illicit crops, and in 2018 the Colombian government under the Duque administration proposed reinstating the aerial fumigation of glyphosate instigating a new round of ongoing legal and political controversy in the country.

Through the ensuing years, as I returned to Putumayo for extended research, to film a popular education project, and to accompany rural communities during the 2013 National Agrarian, Ethnic, and Popular Strike, I was struck not so much by the kinds of violence and environmental destruction produced by the war on drugs, but rather by the tenacity of life in the midst of war. At the margins of the nation's agricultural frontier and experiencing criminalization due to the presence of illicit crops and right- and left-wing paralegal armed groups (referred to as *paracos* and *narcoguerrillas*), the networks of campesinos and indigenous communities I met in and around Putumayo quickly taught me that violence was not the only story to be told. They obliged me to turn my ethnographic attention away from what was raining down on them from crop dusters in the sky to the kinds of propositional life-making processes being actualized in the midst of chemically degraded ecologies. This did not mean that violent death, displacement, and dispossession were somehow less pervasive, but that death was being transformed into something else in the cultivation of gardens, forests, orchards, and ancestral cultivation areas or *chagras*. It was farmers' attunement to the workings of *hojarasca* (litter layers)—the decomposing layers of leaves and stalks that are often used as compost—that led me to rethink the relations between life and death and materiality and politics under everyday conditions of social and armed conflict. It was the potential for hojarasca to "force thought" among rural communities in the Amazon that would become the focus of my fieldwork.

Thus, instead of asking what it means for rural communities to live in coca-growing regions that have been the epicenters of Colombia's geopolitically perpetuated violence, the book that follows this introduction is inspired by practices that render life possible in criminalized and chemically assaulted worlds. How do people keep on and learn to cultivate a garden, care for a forest, or grow food when at any moment a crop duster may pass overhead dousing entire ecosystems with herbicides? Beyond official antidrug policy imperatives to "uproot coca or be uprooted," what other potentialities emerge among rural communities responding to war by cultivating life, which is never altogether separate from death? It was during one of my initial trips to Putumayo that I met an animal husbandry technician and small farmer, Heraldo Vallejo, who would profoundly shape my research questions and ethico-political commitments. Popularly known throughout the region as *el hombre amazónico* (the Amazonian man), I spent the next decade building something akin to what Kim TallBear (2014) calls a "shared conceptual ground" with Heraldo and a dispersed network of alternative agricultural practitioners and organizations throughout the Andean-Amazonian foothills and plains. By a shared conceptual ground, I refer to our attempt to articulate overlapping conceptual and ethical projects while acknowledging our respective situated positions, understandings, and differences as interlocutors and potential collaborators.

As Heraldo explained the situation to me, "The problem in Putumayo is that we do not know where we are standing. This does not have to do with knowledge, but instead with aptitude and attitude. Contrary to what the state says, coca is not our problem. Our urgency has to do with Amazonian agriculture and the displacement and impoverishment of rural families." The networks of campesinos and indigenous communities I came to accompany taught me that "knowing where one is standing" does not refer to *knowing the soil* through a laboratory analysis of its chemical fertility, pH level, or scientific taxonomy, or necessarily knowing any kind of stable entity. Instead, it spoke to the alienation produced by decades of working an export-oriented, monoculture agricultural system—illicit coca being only the last of a protracted series of colonially imposed, extractive-based economic activities that continue to dominate modern territorial relations with the country's western Amazon. Interestingly, when I returned to Colombia in 2008 to begin my long-term fieldwork, the National Geographic Institute Agustín Codazzi (IGAC) launched a natural resources campaign in Bogotá called "The Year of Soils in Colombia." The campaign sought to render soils visible as living worlds with ecosystem services and social functions, and not merely as an exploitable object or resource of political economy. "Place your feet on the soil" was the campaign's 2009 closing motto.

This striking coincidence provoked me to ask: How might techno-scientific calls for one to place their feet on the soil resonate with and diverge from Heraldo's and other farmers' proposal to learn "where one is standing"? In what ways do human–soil relations take on political importance in the complex nexus of antidrug policy, development agendas, agro-environmental sciences, and daily life under military duress? What might the conceptual and material fecundity of soils in tandem with and diverging from more dominant concepts of land and territory teach us? How do soils—what may or may not be conceived of as an object called "soil"—harbor the irreparable wounds and tracks of violence and germinations of transformative proposals and alternative dreams?

This book was written at a time of uncertain transition in Colombia. Certain aspects of the country's more than a half century of social and armed conflict ended in a 2016 peace agreement that was signed between the state and the FARC-EP.

Colombia is engaged in a transitional justice process, with all the risks, expectations, hopes, frustrations, and open-endedness that a post-accord implementation scenario entails. In the chapters that follow, I discuss the ways scientifically informed and non- or *not-only* (as de la Cadena insists; 2015a) scientific practices with "soils" are deeply enmeshed in struggles between farmers, state officials, aid workers, popular agrarian movements, and scientists. These struggles are over the meanings, imaginaries, and material actualizations of "productivity," "rural development," "sustainability," "peace," and what constitutes a "good and just life."

My long-term ethnographic fieldwork was carried out for three years between 2008 and 2011, another consecutive ten months between 2013 and 2014, and varying stints of research before and after this time. I moved between laboratories, government offices, greenhouses, gardens, forests, popular education workshops, and rural mobilizations where I accompanied a heterogeneous group of state soil scientists and technocrats in the capital city of Bogotá and networks of rural social movements and alternative agricultural practitioners in and around Putumayo. Besides attending IGAC's Year of Soils events and seminars, I interviewed a range of scientists, including agrologists, agronomists, soil biologists and microbiologists, chemists, mineralogists, and ecologists. I also conducted fieldwork in the National Soil Science Laboratory of the IGAC and volunteered as a research assistant in the agricultural microbiology laboratory of the National University's Institute of Biotechnology (IBUN) in Bogotá. At the National Laboratory and at IBUN, I assisted in laboratory and greenhouse experiments, took soil microbiology classes, and participated in field inoculations and soil surveys. I was able to learn the practices and fundamental concepts of state soil science and also to pay close attention to the politically charged—often implicit—relationships between soils, land concentration, environmental conservation efforts, territorial conflicts, and the agrarian-based roots of the country's war. In addition to attending national soil science conferences in the cities of Bogotá, Ibagué, and Pereira, I participated in the Eighteenth Latin American Congress of Soil Science in Costa Rica in November 2009. This allowed me to situate Colombian research priorities, environmental legislation, and socio-ecological conflicts within broader scientific debates and policy initiatives across the continent.

Notwithstanding their potentially overlapping conceptual and ethico-political concerns, there are noteworthy differences between rural communities and scientists. Soil scientists' practices tend to take place in laboratories and depend on state research funding cycles, alliances with industrial trade associations, and soil samples transported from rural violence to relative urban safety. Rural communities' practices occur under the militarization of daily life and rely on different kinds of laboriousness and experimental approaches in potentially land-mined gardens, forests, and pastures. The direct relations between the two groups were not predictable, easy, or necessarily even existent. Learning alongside both soil scientists and small farmers quickly complicated any conventional anthropological division between "studying up," in Laura Nader's (1972) now classic sense, to understand the workings of power among experts and institutions, and "studying down" to analyze everyday people's ability to transform and resist these structures.[5] It was impossible to simply oppose "science" and "nonscience," or to assume hierarchical dynamics and fixed locations of subjugation and subversive potential. In this book, I do not simply pit "classic soil science," such as a conception of soil as a reservoir for crop nutrition, against the lively and integral approaches of community agroecological networks and food webs. Nor do I conceive of alternative agriculture as assuming a blanket position against technology or market interactions. Instead, I track how both scientists and rural communities negotiate the boundaries of science and propel their knowledges and practices into political life—if they can—in order to transform the material conditions of different beings and elements that share the contingencies of life and death during protracted years of war.

A growing body of literature at the interstices of feminist and postcolonial science studies and anthropology has made important contributions to our understandings of the intensified shift in the capitalist appropriation of material and immaterial world(s)—or what some scholars have suggested is the commodification of "life itself" (Povinelli 2011a; Rose 2007; Sunder Rajan 2006; Vora 2015). This book both builds upon and departs from this literature by foregrounding the emergence of socio-ecological processes that strive to exist and persist as political, economic, and ethical alternatives to a reductive, market-oriented capture of life. Struggles over these processes lie at the heart of Colombia's ongoing transitional justice scenario as well as the perpetuation of intersecting forms of violence and territorial conflict. They also matter deeply for collective and individual efforts to renew and transform organic and successional forms of life during a humanist capitalist epoch marked by universalizing discourses of the Anthropocene and climate change mitigation strategies that rely on the techno-scientific management of the "environment."

In chapter 1, I introduce my encounter with Heraldo Vallejo in the midst of the US-Colombia war on drugs, and I historically situate the reader within the ongoing extractivist structures that shape territorial relations with and in the country's western Amazon. I begin to familiarize the reader with the urgency expressed by rural communities to cultivate alternatives, what I call selva, agro-life processes. I also outline the ethnographic methods that I engage with to follow human–soil relations, and I present some of the key individual and collective actors that taught me to turn my attention to hojarasca (litter layers).

Chapter 2 focuses on IGAC's 2009 Year of Soils campaign. I discuss the ethico-political implications that the shifting value of soils as living worlds has for the entwined fate of soils and state soil scientists. Taking inspiration from one Colombian scientist, Abdón Cortés, and his creative notion of soils as *el teatro de la vida* (the theater of life), I speculatively imagine what I call the "poetics of the politics of soil health" by allying scientific conceptualizations and poetic forms of soil sensing.

Through ethnographic engagement in laboratories and on state soil survey trips and small farms, chapter 3 places scientific and campesino practices in conversation. I expand upon a concept I borrow from Heraldo, cultivating *ojos para ella* (eyes for her)—*la selva*—to demonstrate the partially coinciding, diverging, and incommensurable relations that emerge between caring for the soil for the purpose of scientific interests and economic imperatives and caring for a world full of beings that mutually nourish each other.

In chapter 4, I follow the diverse material practices and corresponding life philosophies of a dispersed network of rural families and alternative agricultural practitioners throughout the Andean-Amazonian foothills and plains. Rather than productivity, one of the central elements of modern capitalist growth, I explore how the regenerative capacity of the selva relies on organic decay, impermanence, decomposition, and even fragility that complicates modernist, biopolitically oriented bifurcations of living and dying.

In chapter 5, I think with Brazilian-based agronomist Ana Primavesi's agroecological term *espacios vitales* (vital spaces) as a conceptual and political tool from which to imagine what an Andean-Amazonian territory of what I call "resonating farms" might look and sound like. This chapter focuses on lessons learned through my conversations with Heraldo Vallejo, specifically on the kinds of conceptual personage and *pensamiento propio* (a group or collective's own thinking) that constitute selva agro-life processes. Chapter 6 expands upon the kinds of potentialities and limits that arise from different sets of human–soil relations, relations that may unravel or destabilize concepts of the human and of soil, and of their hyphenated pairing. I return to the material and conceptual fecundity of hojarasca to discuss what can be learned from transitional states rather than stable entities and from "soils" that never became the industrialized or chemical soil that has taken center stage in global concerns over anthropogenic climate change.

While selva is often translated into English as "jungle"—a word imbued with a complex colonial history—I continue to use the Spanish word selva at specific moments throughout this book. This is because I learned with rural communities to treat selva as a concept, a relational set of practices, and an existent or living force rather than an entity that can easily be divided into units or reduced to a representational landscape descriptor. These communities explained to me that the word *bosque* (forest) does not necessarily convey conspicuous, complex biodiversity since it may refer to a monocultural system of trees or a collection of commercial timber–yielding varieties destined for extraction. Selva also works to resignify *monte* (brush, forested hilly land) and *rastrojo* (animal fodder, weedy regrowth) when the latter are used disparagingly to describe an unruly and/or uninhabited landscape that should be cleared, feared, or tamed.

The playful and politically motivated conceptual work of the rural practitioners I accompanied and thought with led me to ask what forms of writing are necessary to articulate a selva analytic. By this I refer to an analytic that aspires not only to write about the selva or like a selva, but instead to follow and perform the shifting relations, temporalities, and material and immaterial textures that compose and decompose selva life and death. The question of how forms are tied to, constitutive of, and transmitted via other than humans, species, beings, and elements has received growing ethnographic attention in McLean's work on poetics and materiality (2009), Choy's (2011) account of the "four forms of air," Kohn's (2013) work on forest semiosis, and Myers's (2015a) and Hartigan's (2017) respective explorations of plant sensorium and intelligence, among others. I conceive of writing selva not as a romantic mode of "giving voice" to a tropical forest ecology, but rather as an attempt to attune to and perform selva modes of expression that may work through literary and poetic genres that hold geologic, human, and microbial temporalities in tension and simultaneity with the analytical languages of the social and ecological sciences. I use vignettes and poetic forms of writing at different moments to interrupt the narrative, which is primarily focused on the cultivation of life in the midst of death. These interruptions attend to the acts and latent threats of violence that explode into the everyday, producing gray spaces between official times of war and transitional periods of peace. Territorial conflicts and structural violence was and continues to be a destabilizing condition of daily life in rural Colombia, and it shapes the material sediments and embodied memories harbored in and expressed by regional ecologies even when it was not the explicit focus of my interlocutors.

1

FROM AERIAL SPACES TO LITTER LAYERS

PULSATION

Having momentarily broken away from the group, I stood, waiting in silence, between rows of climbing plants, fruit trees, beds of tubers, and shrubs. The sunflowers by my side leaned forward only slightly. A nearly imperceptible breeze caused the vines to quiver. The scent of rotting fruit rinds rose up from the ground. There was a distant flap of wings. Much closer I could hear larvae chewing on the *granadilla* leaves and insects humming from within knotty bundles and flower petals. It was much cooler here, and after a few seconds the humming grew more intense. My only words for it: a hundred damp index fingers gliding around the rims of water glasses. Life as it pulsates, withers, draws at once a next and last breath. This was the distinct sound—better yet, force—permeating the air when I first stepped off the bus in San Miguel, Putumayo, at the Amazonian farm school La Hojarasca, which means "litter layer" or, more colloquially, decomposing leaves often used as compost. I looked around, mistakenly surveying the landscape for some discernible source for the hum. Groups of campesinos were conversing next to a wooden table lined with seeds as big as fists, others as tiny as mites. Hens and wild turkeys roamed about, trampling through the underbrush. A family of geese honked noisily as they descended upon their lunch of minced sugarcane. There was the crunch of human feet pressing down on layers of dry leaves and stalks, a sound quite distinct from the squish of boots sliding against bare, clay-laden mud. I heard the slice of a machete, the heavy thump of *copoazú* when the fruit hit the ground, laughter, more buzzing, and the friction of scraped stone as *bore* leaves were ground into grain. The trees were lined with *mochilero* nests, and every so often I could make out the birds' calls—the sound of water, the reverberation of a drop of water, that almost electric sound it makes in the exact moment it hits a surface and morphs into disparate forms. I was so absorbed by all that was going on around me, all that it was doing and undoing in me, that I failed to sense another human presence until I heard a voice call out from behind. "Wouldn't you agree," the person asked as he approached where I stood in the middle of the creeping plant garden, "life makes life happier?"

It was anything but a simple question. This was my first encounter with Heraldo Vallejo and the powerful yet vulnerable force emanating from a collective proposal such as La Hojarasca.

"STRAPPING ONESELF TO THE TREE WITH THE MOST SHADE"

An incandescent wooden cross
A blackboard littered with bullet holes
A calloused hand cradles a half-rotten pineapple . . .

When I met Heraldo I had spent the previous week on a Witness for Peace delegation consisting of a mix of college students, congressional aids, lawyers, and activists from the United States. As the nonprofit organization's name suggests, we traveled to Putumayo in August 2007 to "witness" the adverse impacts of Plan Colombia on local communities and landscapes and to gather testimonial and other kinds of evidence for international advocacy efforts opposing the US foreign policy.[1] The violence wrought by war creates its own humanitarian regime, a dialectical and denunciatory shadow that at many times during the trip seemed ironically to me to be the opposite side of the same militarized coin. The day before our visit to La Hojarasca we accompanied a woman in a field of dying cacao plants that were part of a USAID illicit crop substitution pact communities had signed, and that had been fumigated two weeks earlier. "There is no way to say this," she told us. We stood for a long while in her field in silence. A group of indignant neighbors gathered near the front of the house, displaying deformed plantains, rotting pineapples, wilted husks, and more cacao leaves riddled with holes and spots. They expressed uncertainties about the protracted life span of the herbicide glyphosate in human bloodstreams, soils, and watersheds. They recounted the hens that were stolen, the number of fences knocked down, and the verbal and physical abuse they suffered when forced manual eradicators, accompanied by military and police, ripped out whatever stubborn coca plants were salvaged from the crop-

FIGURE 1.1 Mochilero nests in San Miguel, Putumayo, August 2007. Photograph by author.

duster planes by equally stubborn and artful farmers. Rows of gaping holes in the ground, unemployment, hunger, and human rights violations were left behind in the eradicators' wake.

Rural families, whether they had been farmers before commercial coca gained momentum or ended up migrating to the region and working in different capacities that integrated them into the coca economy, had serious doubts whether it was even possible to sustain agricultural livelihoods in Putumayo—coca or otherwise. Much of the countryside was haunted by an eerie silence. Many people, including an agronomist employed by the provin-

FIGURE 1.2 Plantain aerially fumigated with glyphosate in Valle del Guamez, Putumayo, August 2007. Photograph by author.

cial secretary of agriculture, told us that they refused to grow any more food and commercial crops until the state and/or US embassy definitively abolished aerial spraying. Antidrug policy *"está acabando con la vida"* (is finishing off life), they told us. Toxicity seemed likely. One waited to be deprived of the resources that allow living beings to thrive. At any moment the sting of a droplet, dampened leaf, inhibited enzymes, and the end to synthesis—a strangling of life from the inside out.[2] It was not lost on any of them that their life staple had been excised so that life elsewhere, and in the wounded soil itself, could claim to be secured, sanitized, protected, and flourish.

Communities that agreed to self-eradicate their coca as a prerequisite to participating in USAID alternative development programs waited for the next round of shifting subcontractors infamous for their systematic mismanagement of project budgets. The impromptu community organizations that assembled to ensure funding allocation were often disbanded as soon as the project cycles ended. People had become mostly concerned with how to enroll in the next state aid program or with, in the words of one campesino, "how to strap themselves to the tree with the most shade." During my interviews at the United Nations Office on Drugs and Crime (UNODC) in Bogotá, which houses the System for the Integral Monitoring of Illicit Crops (SIMCI), officials expressed doubts about why the state should be obligated to develop remote areas of the country where individuals had penetrated, indeed deforested, to more covertly engage in illicit activities. Rural families, on the other hand, argue that they have been obliged to migrate to marginal zones by historic cycles of violence and dispossession in the country's agricultural interior, the structural lack of access to viable markets and state services, the implementation of neoliberal policies that have worsened urban and rural poverty, and the repressive and indiscriminate nature of antinarcotics policy itself. During Phases I and II of Plan Colombia (roughly 1999–2010), alternative development paradigms transitioned from an initial focus on social pacts and crop substitution, to creating a "culture of legal economic practices" through agro-industrial–oriented projects and partnerships with private enterprise. Assistance-based interventions, which have always been conditioned upon a "zero tolerance/zero coca" imperative, relegated rural communities to the role of beneficiaries. Alternative development has tended to follow the same market logic as commercial coca, aiming to replace an illicit export-oriented crop with legal cash crop varieties, such as black pepper, coffee, vanilla, heart of palm, heliconia flowers, and cacao.

In my interviews with USAID employees, they explained more than a decade and $80 million of failed development projects in Putumayo as "an unfortunate but instructive learning curve." The cost of production in remote areas with little infrastructure was much higher than anticipated. Market studies were not conducted to ensure a niche for newly introduced agro-industrial products. Offering aid only to coca-growing families stimulated the planting of coca and failed to help individuals who did not grow the plant but were employed within its broader commodity chain. Market-oriented crops were plagued with problems of "quality control" (i.e., rampant fungus and tropical pests). The expectation that people would snitch on their neighbors and forcibly rip out their coca plants only served to fracture community relations and worsen social conflicts. This list did not include the other USAID fiascos that rural communities attested to: the delivery of beakless hens imported from the United States that were so useless they went straight into a boiling pot for lunch; cows that were distributed to families without access to pasture land ended up being sold to narcotraffickers; a meat-processing plant was shut down indefinitely after FARC-EP guerrillas blew up the regional power generator; administrative corruption bankrupted the heart of palm factory three times; pepper farmers were now defaulting on their loans because the pepper took six months longer to mature than agronomists had advised; heliconia

FIGURE 1.3 Aftermath of aerial fumigation with glyphosate in Bajo Putumayo, August 2007. Photograph by author.

flowers destined for Bogotá supermarkets were home to a maggot that was now attacking the local varieties of subsistence plantains.[3] The list goes on, and the level of tragic absurdity, generalized incompetence, and economic squandering begins to sound conspiratorial.

Our delegation visited La Hojarasca farm school after witnessing a series of these failed USAID projects. The school was tucked around the corner from a field where a large wooden cross stood solemnly behind an abandoned windowless house. The house had served as a right-wing paramilitary interrogation center where victims were tortured and disappeared between 1998 and 2006. During this time, the coalition paramilitary organization, the United Self-Defense Forces of Colombia (AUC), occupied the urban centers of the subregion known as Bajo Putumayo and disputed with FARC-EP guerrillas for control over the local population, territory, and taxation of the cocaine trade.[4] The cross was erected to mark the presence of a mass grave that could not be officially disclosed to local authorities given the ongoing nature of war and state complicity in paramilitary violence. A forested brush line that survived confrontations between the armed groups ended up serving as a "natural" barrier between the FARC, who were relegated to rural corridors, and the AUC, which assumed control over the town.

Don Pedro, a student at La Hojarasca who lived nearby with his oldest son and two grandsons, trekked along with us on the muddy branch road that rural residents, with the aid of the FARC-EP, carved out of surrounding forest in the 1980s to provide more or less traversable infrastructure to the municipal town center. All of Putumayo lacks access to potable water, and Don Pedro's community remains without electricity and basic sanitation, like so many other rural areas of the country. We passed a towering *ceiba* tree bordering his landholding that had been misted with glyphosate. With almost strange precision the tree appeared to have divided in two. One half slowly deteriorated while the other seemed to be resisting death, unremittingly vibrant and green. The ceiba had been there as long as Don Pedro could remember, and he hoped that the remaining living half would endure long enough to attract pollinating bats and offer up seeds to be dispersed by the breeze. La Hojarasca, he explained, was part of a larger sustainable development project called San Miguel Mira hacia Colombia y el Mundo (San Miguel Looks toward Colombia and the World), coordinated by the Jesuit-run NGO Center for Research and Popular Education (CINEP), based in Bogotá. During the two years that it functioned as a school before the funding cycle ended and CINEP withdrew its projects from Putumayo, around two hundred farmers from three of the department's thirteen municipalities graduated with a diploma in sustainable Amazonian agriculture.[5] Heraldo Vallejo, who was fifty years old at the time we met, was one of the central intellectual contributors to the design and conceptual foundation of the farm: in particular, the *campesino a campesino, indígena a indígena* methodology that aspired to multiply what I would later hear referred to in heterogeneous ways as Amazonian, analogous, successional, biological, mystical, and selva agricultures.

LA HOJARASCA

Indeed, what most impacted me when I first arrived at La Hojarasca was the pulsation the farm school was generating, literally resonating by its very existence in the midst of a criminalized ecology. It was bundles of life pulsating away—dense entanglements of diverse plants, decomposing leaves and rootlets, the buzz of insects, sounds of small animals and birds cloaked by selva canopy, and human activity—that allowed it to carve out a transformative space for itself, even if precariously so. This was not a space where life was simply enduring within social suffering, abandonment, and contamination, but one where other modes of eating, growing, seeing, walking, exchanging, cultivating, composing, and hence decomposing were being set in motion. The farm school, in Michel Serres's terms, was a kind of *static* that garnered power not because it occupied or emanated from any center, but because its germination temporarily filled the surrounding milieu (2007, 95). For Serres, static acts as a kind of interference or noise that, while disabling certain connections and pathways of communication, also facilitates others. Similar to his grasshoppers in *The Parasite* that keep on singing and filling a space in their counterattack against the noise of chainsaw motors that attempt to cover and displace them from the forest, smaller pulsations of energy emanating from the farm school were chasing out bigger and more devastating ones. In that moment, life was making life happier. The farm seemed to me a grasshopper noisily resonating and defending a space for itself by interrupting the expansive silencing and destructive mechanisms of a demoralizing and repressive war on drugs.

Once La Hojarasca stopped receiving CINEP funds, however, it transformed into something else: no longer a school, but a farm returned to its original owners, the Agro-Industrial Integral Association of Campesinos in San Miguel (AAINCOS), to be managed by a single family until further funding could be secured for a new round of students. It was unclear if the school would carry on. In this sense, it did not depart from a history of provisional, externally funded development initiatives in Putumayo. The school was not rooted in fixed conditions of possibility, but something closer to what Kathleen Stewart refers to as the "actual lines of potential that a *something* coming together calls to mind and sets in motion" (2007, 2). In dramatic contrast to the neighboring USAID programs, however, which parachuted down from "center to periphery," La Hojarasca attempted to gather together and multiply diverse aptitudes, desires to learn, and already existing modes of relating to selva or Amazonian agriculture. The farm school, not as a delimited project with a calendared timeline and externally determined budget, but rather as an ongoing *process*, was capable of picking up the kinds of affective densities and textures that Stewart conceptualizes in terms of the relational thickness of the present. This thickness is inhabited by the actual thoughts, feelings, dreams, ethos, and material transformations that exist and are made possible in processes striving to shake loose, to whatever degree, from dominant definitions in order to become something different.

La Hojarasca was propelled by the desire of a growing number of rural communities to distance themselves, even if only modestly, from the moralizing discourses and stigmatized overcoding of existing state categories: *cocalero* (coca grower), *auxiliador de la guerrilla* (guerrilla helper), *beneficiario del desarrollo* (beneficiary of development), *mendigo* (beggar), *víctima* (victim), *colono desarraigado y depredador* (rootless and predatory settler), *población flotante* (floating population), and even Putumayo campesino. The latter was a category that I found myself and oftentimes them using, which reminded us both just how difficult it was to escape the ensnaring of state-codified recognition. More than simply denouncing or resignifying these disputed identities, the visit to the farm school was my first lesson in learning that distancing was contingent upon a whole series of material, conceptual, and ethico-political transformations in and with a particular Andean-Amazonian ecology.

Some of the campesinos I met that day had rejected commercial coca since its arrival in Putumayo, even though they respect and grow coca plants as medicine, sustenance, a spiritual force, and a constitutive element of local biodiversity. Others were cocaleros actively seeking a transition out of the coca economy, worn down by state persecution, fluctuating market prices, rising costs of production, and the untenable nature of a monocrop agricultural system in the Amazon.[6] Monoculture coca initiated the arrival of agrichemicals in the territory in the late 1970s and led to a reduction in the diversity of seeds being sown. This profoundly altered and homogenized the varieties of local seeds, plants, trees, and the range of recipes and eating practices, largely displacing subsistence and commercial food production. When the FARC-EP enforced an armed strike in 1998 that cut Putumayo off from the country's Andean interior and blocked the transport of foodstuffs, rural communities became more acutely aware of their mounting food, and hence economic and political, dependencies.

However, as the students at La Hojarasca explained, there was no simple or romantic return to the days before coca when many rural families engaged in unsustainable agricultural practices that they imported from their Andean places of origin. They referred to the decades of *tumbando bosque* (clearing forest) to sow rice and corn. These lead to quickly waning harvests after naked soils are exposed to intense tropical sunlight, heavy rains, and nutrient extraction without the protection of selva canopy and the return inputs of decomposing hojarasca. Perceived to be degraded soils, these plots are then converted into open pasture and inevitably sold to individuals with capital (i.e., cattle). This leads families to clear more forest, only to repeat the vicious cycle. Rural families keep finding themselves displaced deeper into the selva and farther away from town markets and access to transportation. Their costs of production increase and are never compensated by the meager prices paid by intermediaries for traditional subsistence crops like plantains, bananas, and yucca. In the 1970s and 1980s, the State Institute of Agriculture and Livestock Merchandising (IDEMA) purchased rice and corn from campesinos in frontier regions like Putumayo to sell at subsidized rates in the low-income urban neighborhoods of the country's major cities, such as Bogotá, Medellín, and Cali. Rural families remember their experience with IDEMA as one in which they were often left in the lurch, waiting for days to get paid, skulking around town, sleeping in public parks, and wasting time and money far

FIGURE 1.4 Learning with the creeping plant garden and Heraldo Vallejo. Mocoa, Putumayo, January 2013. Photograph by author.

from their farms. For many, these farms are hours away by a combination of travel on foot, canoe, and mule or horse ride. After IDEMA was shut down, Putumayo's Livestock Fund (Fondo Ganadero) eventually went bankrupt due to corruption. Soon after, commercial coca plants arrived in the region, offering a relatively higher income, easy transport, and guaranteed market despite the heavy investments that the monocrop requires in chemical pesticides, fertilizers, and other assorted chemicals when people also engage in the making of basic cocaine paste.

The campesinos I met at La Hojarasca, as well as the rural families, networks, collectives, and agrarian social movements I would come to accompany over the following years, explained that they began to view both monoculture coca and its official alternatives as forcing not only people, but also the life forms with which they live and labor, into capitalist relations of production. They critique this singular economic logic, in part, because it treats commodity exchange as the founding principle of agri-sociality. Underpinning this sociality is an assumption that campesinos are simply poor farmers lacking the proper technology, financial capital, business alliances, and work ethic to become full-blown agro-industrial farmers. Nelso and María Elva, a campesina family that was displaced to Mocoa and with whom I built a deep relationship, characterize this as "campesinos playing capitalist without capital." They argue that this economic model dismisses small farming as an obsolete livelihood of the past. It has also turned Putumayo into a laboratory for public policies gone awry by denying the cultural particularities and diversity of campesino families and economies, as well as the agroecological conditions of an Amazonian territory. These families do not reject all market transactions, but rather the reductionist and inevitable way "the market" has come to be delimited as capitalist, export-oriented, industrialized, competitive, and solely commodity- and cash-based. While diverse modes of production, exchange, and labor throughout the region have, of course, not been entirely eliminated, *trueque* (barter), *mingas* (communal labor, cooperative work), *mercados campesinos* (campesino markets), and other non- and anticapitalist economic relations have ostensibly been marginalized.

However, as I mentioned in the introduction, the ongoing presence and forty-year legacy of monoculture coca production was not portrayed as the only or even the primary "problem to be solved." In a Stengerian sense, the concrete situation—better yet, la selva—had come to oblige a different politics of attention and ethics of response. Stengers invites us to expand the scope of our understanding of obligation when she equates being obligated by a situation as giving the situation the power to make us think (2005c, 185). In this sense, the formulation of a problem can never be dissociated from its *oikus*, that is, from a milieu or environment that requires a specific ethos and analytical engagement. As we sat conversing in the open-air classroom of La Hojarasca, Heraldo explained to me that to historically contingent and differential extents, many indigenous and nonindigenous rural inhabitants, both those who settled in the region and those born in Putumayo, were explicitly taught or inadvertently ended up "not knowing where they are standing."[7] Not knowing where one is standing is a result of structurally produced forms of estrangement that alienate people from the myriad relations that compose and decompose the place in which they are physically located, and thus from the world of which they form a part—a world, I would come to learn, where cultivating entails processes of cultivating and being cultivated by what I describe in subsequent chapters as ojos para ella (eyes for her), la selva.

EL HOMBRE AMAZÓNICO

Born in the rural *vereda* (rural district) of El Pepino, on the outskirts of Putumayo's capital, Mocoa, in 1957, Heraldo spent most of his childhood farther south after his grandfather convinced the family to move to the municipality of Puerto Asís, where land was more accessible. He worked tending livestock, cleaning pastures, selling cheese and milk, and harvesting rice and pineapples while growing up, "like any other campesino," as he put it. The money he earned went to pay for his school expenses since his parents could not afford to educate their five children. Even then, he was forced to drop out of high school and later made up the missed year. Heraldo was one of only two students from his rural elementary school with the resources to go on to study at the closest university, in the neighboring Andean state of Nariño. Following imprisonment for six months as a Marxist-inspired leftist student leader, and after his graduation with a degree in animal husbandry, Heraldo returned to Putumayo in the early 1980s and began, as he says, to "unlearn" the dominant teachings of the modern agricultural and veterinary sciences.

This process of unlearning, and hence relearning, was propelled by the fact that Heraldo, as he said, "had always witnessed other ways of doing things" on the part of indigenous and campesino neighbors and some of his family members. Upon returning to Putumayo during a decade when the monoculture coca boom was gaining momentum and the first alternative development and crop-substitution projects were implemented and failing, Heraldo began to question the applicability of dominant agricultural models in the Amazon as well as the underpinning productivist logics informing these models. As I return to in chapter 4, Nelso and María Elva cite the influential outreach work of the Catholic liberation theology–inspired priest Padre Alcides Jiménez for motivating the first Amazonian-oriented agricultural alternatives after the arrival and expansion of monoculture coca crops. During the 1980s and 1990s, Alcides was well known for distributing traditional Amazonian seeds instead of the host during the masses he led at Nuestra Señora de Carmen church in Puerto Caicedo, Putumayo. Heraldo uses the term *agricultura de la muerte* (agriculture of death) to refer to the extractive-based practices that result when rural families perceive themselves as external to, rather than shaping relations within, the cyclical conditions of existence of la selva. More concretely, agriculture of death refers to reliance on chemical inputs, the introduction of patented and transgenic seeds, monoculture and export-oriented systems, land titling premised on deforestation, Green Revolution technology transfers, and more recent neoliberal reforms—all of which are perceived as strangling out and rendering obsolete diverse campesino and indigenous practices and economies.[8] Anthropologist Henry Salgado Ruiz employs the term *necro-política agraria* (agrarian necropolitics) (2012, 4) to describe the war waged against campesinos in Colombia since the 1920s. He uses the term to critique the concept of "spontaneous colonization" that is often used to characterize the settling of the country's frontier territories, which he argues obscures the historical and ongoing processes of exclusion and dispossession of rural communities as well as the dominant political relations and structural violence shaping the country's agrarian sector.

Given the historic lack of agroecologically appropriate state-based technical assistance in Putumayo, the decades of breached political agreements on the part of the national government after civilian strikes, and the ways that both monocrop coca and repressive antidrug policy have largely eliminated local food production, Heraldo and a small number of campesinos have come to provide an informal network of regional alternative agricultural technical assistance. Over the years, Heraldo became popularly known as el hombre amazónico (the Amazonian man) among campesino, indigenous, and Afro-Colombian communities, as well as certain state technocrats and bureaucrats. This name is an acknowledgment of his Amazonian life philosophies, critical ethos, and practical technical orientation. His reputation and trajectory led to his formative role at La Hojarasca. After our meeting at the farm school, he invited me to visit his own two-hectare farm outside Mocoa, which he is similarly working to convert into a collective learning and gathering space. Even though Heraldo is highly regarded by diverse sectors, at times I heard state functionaries refer to him as a "visionary utopian" or "dreamer" that Putumayo was not yet ready to embrace. In worst-case scenarios, his expertise was discredited because an animal husbandry technician's compartmentalized knowledge was said to be limited to animals and not agriculture. Even more dangerously, his ideas were at times said to be blatantly "communist." In a country where "communists" have been systematically persecuted and assassinated, this kind of stigmatizing label is no small threat.

During my long-term fieldwork, I joined Heraldo and other alternative agricultural practitioners when they worked in solidarity with campesino associations, unions, and indigenous communities attempting to transition from monocrop coca, official alternative development programs, and other unsustainable agricultural systems toward a heterogeneous range of selva agricultures. This included attending meetings and protests with the Mesa Regional de Organizaciones Sociales del Putumayo, Baja Bota Caucana y Cofanía Jardines de Sucumbíos, Nariño (MEROS); popular education workshops with Inga and Nasa indigenous reservations; seed fairs of the Network of Guardians of the Seeds of Life; the mingas of communities affiliated with the legacy of Padre Alcides's Nuevo Milenio initiative and ecological evangelism in Puerto Caicedo; and workshops with rural communities participating in the Environmental Clinic in neighboring Sucumbíos, Ecuador.[9]

MEROS is a regional network of around thirty-five social organizations that publicly regrouped in 2006 after the intense decade of paramilitary repression against popular leaders and movements. Around 80 percent of the organizations self-identify as campesina, and they inherit different trajectories of political protest, community leadership, human rights defense, and organizational work on the part of rural laborers, coca growers, campesinos, workers in the oil industry, civic organizations, indigenous and Afro-Colombian communities, and youth and women's organizations dating back to the 1970s.[10] Heraldo began to collaborate with MEROS in 2008 after building trust with its member organizations and overcoming their deep wariness of technicians as well as the political interference of nongovernmental intermediaries.[11] Heraldo, and I to a lesser extent, collaborated on the design of a proposal for the formulation of what MEROS calls the Andean-Amazonian Integral Development Plan (PLADIA 2035). PLADIA is the current iteration of the community-based development plan that was first proposed during negotiations with the state after the massive *Marchas Cocaleras* (coca growers' protests) in southwestern Colombia in 1996 (see Ramírez 2001). Despite written and verbal agreements made at the time, the precursor to PLADIA was not supported by the state, and instead became one of at least twenty-eight accords that were signed between the national government and regional social movements over the last twenty-five years that have not been upheld (MEROS 2015, 10). Social organizations participating in the 1996 strike aspired to provide an alternative to militarized antidrug policy and failed state-led development models by addressing the structural conditions that lead to illicit coca production and the displacement and impoverishment of rural families.

More than a development model, PLADIA is an integral life plan that proposes to reconvert the region's economy by "dreaming with a Putumayo different from what neoliberalism offers [us]" (MEROS 2015, 14). MEROS conceptualizes the formulation of PLADIA as a radical participatory methodology for popular territorial planning in which rural communities un- and relearn the agroecological and sociocultural realities of an Amazonian territory. This includes reimagining housing, education, health care, recreation, infrastructure, and the region's agro-environmental systems while also recovering and revindicating traditional and ancestral *saberes y sabores* (know-how/wisdom and tastes/eating). I was invited to join MEROS's technical team during a different period of strike negotiations with the government in 2013–14 and to facilitate popular education workshops with urban and rural communities throughout Putumayo. The 2013 National Agrarian, Ethnic, and Popular Strike was primarily organized by campesino movements and diverse agrarian associations, small-scale miners, and health, education, and transportation sectors in twenty nodes throughout the country. The strike called for the suspension of the free trade agreement with the United States and structural transformations to deal with the country's multilayered agrarian crisis; the participation of small miners in mining policy and an end to a national development model fueled by industrial extractivism; recognition of the political and territorial rights of rural communities; alternative legislation to combat the increasing privatization of health and education; and a reduction in the rising costs of transportation and fuel.[12] During this time, I learned the material, ethical, and conceptual components of PLADIA's proposal to transition to *fincas agroproductivas sostenibles* (sustainable agro-productive farms), and how these farms differ from and enter into dialogue with ancestral cultivation areas or chagras. In 2016, the community-based diagnostic stage of PLADIA was funded by the government after a new round of popular protests in Putumayo, but the formulation and implementation of the integral life plan is still pending.[13]

Without losing sight of the high incidence of informal claims to land rights in Putumayo, there is a relatively democratic distribution of land if one takes into consideration the government-established family agricultural unit, known as the UAF. In theory, the UAF is the extension of land assigned to a campesino family by the state in land restitution projects and in the titling of *baldíos* (deemed uncultivated land). The UAF extension is determined based on conventional paradigms of soil fertility and the productive projects and technologies of each farm in a specific zone. It is intended to limit individual land ownership and to implement a system of "ordered distribution and rational land use" throughout the country (Law 160 of 1994, article 1). The UAF in Putumayo oscillates between 10 to 45 hectares in the conventional "highly fertile" soils of the Andean subregion known as Alto Putumayo; 35 to 45 hectares in the "less fertile" Andean-Amazonian foothills; 70 to 120 hectares in the "infertile" Amazonian plains or what is referred to as Bajo Putumayo;

and 212 to 286 hectares in the lower part of Puerto Leguízamo from Puerto Ospina up to the border with the neighboring department of Amazonas (Centro Nacional de Memoria Histórica 2015, 48).

As when I met Heraldo in 2007, he continues to be periodically contracted as a freelance technical consultant by different government institutions and nonprofit organizations. During my fieldwork these included contracts with Parques Naturales Nacionales de Colombia (National Parks Service), CORPOAMAZONIA (the regional environmental authority), and the provincial secretary of agriculture. Heraldo accepts these contracts when they enable him to share his Amazonian technical orientation and corresponding life philosophies with rural communities, and also because the salary he earns is directly invested into the conversion of his farm into a farm school. This said, he turns down jobs or resigns from positions that he perceives to be contradictory to what I call his *Amazonian agro-life process*. For example, in the early 1990s and then again in 2016, certain electoral openings and political shifts occurred in the administration of Putumayo, and Heraldo was appointed as the provincial secretary of agriculture.[14] On both occasions, he resigned from the post due to the levels of corruption and lack of political effectiveness, which he argued obstructed his ability to implement agroecologically appropriate and just agricultural policies that would benefit the region's rural sectors, which constitute over half of the population. The majority of Putumayo's rural residents do not hold land titles, and around 77 percent live and work on small to medium parcels that have an area of less than one hundred hectares. It is important to note that during the time of my fieldwork, Putumayo's secretaries of health and social protection and education interfered with the national ministries due to mismanagement. Two democratically elected governors were deposed on corruption charges, and four more interim governors were appointed over a five-year period, each one designating a different set of secretaries and cabinet members. For two years during what was referred to in Putumayo as a "crisis of governability," I had the opportunity to accompany Heraldo while he engaged in his life work. This included farm work, his jobs as a contracted employee with state institutions and nonprofit organizations, and his efforts to build solidarity networks with dispersed Amazonian agro-life processes as local farmer, alternative technician, and hombre amazónico. In diverse ways, these processes share a desire to actualize alternatives to the protracted history and ongoing extractivist practices and enclave economies that continue to characterize modern territorial relations with the country's western Amazon.

FIGURE 1.5 Traditional horse path in Putumayo. Photograph by author.

This history of violent displacement and territorial reconfigurations is crucial for understanding the socio-environmental contexts, political urgency, and material conditions out of which Amazonian agro-life processes emerge and respond to.

VIOLENCE, EXTRACTION, AND COLONIZING STRUCTURES

Putumayo is one of six states and three transitional zones making up what is known as Colombia's Amazonia, a region that constitutes around 40 percent of the national territory. Derived from Quechua, Putumayo roughly translates as "river that flows until where the sun rises." Spanish conquerors first invaded the territory in 1538, followed by the arrival of Jesuit and Franciscan missionaries throughout the sixteenth and seventeenth centuries, sent to consolidate ecclesiastical jurisdiction and to evangelize and consolidate the territories' diverse Native peoples. These missionaries were forced to periodically enter and abandon their colonial townships due to continuous attacks organized by "rebellious" *piedemonte amazónico* indigenous groups who lived dispersed yet economically and culturally integrated throughout the selva and foothills of the Andes. Whenever I rode the bus past Pueblo Viejo, the site of the original colonial construction of Putumayo's capital, I heard stories about how Andakíes burnt down the burgeoning city four times between the 1500s and 1700s. Andakíes are said to have negotiated alliances with other Andean-Amazonian peoples, such as the Tamas, Sucumbíos, Mocoas, Inganos, and Sibundoyes, to defend their territories—gateway to the expansive selva—against the Spanish Crown. In 1784, Franciscan missionaries wrote of "savage infidels," *aucas* or *indios no bautizados*, under the direction of powerful *chamánes* who ingested *yagé* and transformed themselves into pumas that devoured entire villages of Christians (Ramírez 1996, 90–92). One can imagine the fearful and furtive glances of priests as they peered out their windows and down torchlit paths waiting for the vengeful pumas to pounce.

The Andakíes, like all other "rebellious tribes," are said to have fought to their collective death. Yet Heraldo told me a different story, in which after their last attack against the city, the Andakíes rendered themselves invisible to the missionaries and their accompanying military entourage rather than succumb to colonization or, what is the same, annihilation. Since then, the Andakíes continue to roam freely throughout the Amazonian foothills and plains. From helicopters hovering above remote oil concessions, company technicians claim to have seen campfires and smoke, but never any people or footprints. For Heraldo, Andakíes, *indios-aucas*, continue to resist, only revealing themselves to master chamánes or *taitas* with the help of yagé, a plant-based hallucinogen taken in certain indigenous ceremonies, when they impart the secrets of the selva. Perhaps the oil technicians saw the stirrings of pumas-aucas who tread lightly and disappear into the selva with a fleeting swish of their tails. In chapter 5, I return to these pumas-aucas when I discuss modes of decomposing into selva life and the politics of imperceptibility. When the first census was conducted in 1849, the local population, considered only to be those living in the existing twenty colonial settlements, was categorized into "*racionales*" and "*indígenas civilizados*." The rest of the territory was actively imagined as uncultivated and uninhabited, that is, *territorio baldío*, systematically emptying the selva of the existence of its millennial inhabitants and rendering the territory a receptor of future waves of settlers and displaced populations from the country's Andean interior.

Anthropologist María Clemencia Ramírez (2001) characterizes five waves of colonization in Putumayo strongly tied to extractivism, commencing with quina and rubber, then followed by gold, timber, especially *cedro* trees, and otter, tiger cat, black caiman, and musk hog skins during the first half of the twentieth century. While the rubber industry failed to generate permanent settlements, it initiated the expansion of the nation's agricultural frontier, and produced the violent genocide and further dispossession of indigenous peoples chronicled by a number of historians and anthropologists (Ariza, Ramírez, and Vega 1998; Bonilla 1969; Taussig 1984). The first roads initiated by missionaries were completed thirty years later during the Colombia-Peru War (1932–33) when nationalist sentiments were sparked toward the Amazon in reaction to Peru's increasing encroachment on Colombian territory (Palacio 2004; Santoyo 2002). Some of the rural families I came to know shared stories of how their parents had been recruited from neighboring regions to defend the sovereignty of the country's southernmost border and to construct the first military bases and motor vehicle–traversable roads connecting the western Amazon to the Andes. Early military colonization (Culma 2013), and the arrival of settlers from Nariño to pan for gold, was followed by waves of rural families displaced from the nation's Andean interior. A long period of violence between Liberal and Conservative parties, known as La Violencia (1948–60) engulfed the country and led to the creation of the first liberal guerrilla movements and armed conservative mercenaries, or *chulavitas*. Putumayo was a recipient of campesino families and indigenous communities fleeing this violence and whose access to land had been limited by expanding latifundia. The dissolution of indigenous *resguardos* in Nariño in the 1940s, and the generalized and multitiered violent pressures on indigenous and campesino lands led Pasto, Embera-Katío, Embera-Chamí, Nasa, Awá, Korebaju, Murui, Huitoto, Boras, Kichwas, and Guambiano peoples to move into Putumayo, further impacting the territories and movements of the officially recognized ancestral Kofán, Siona, Kamsá, and Inga peoples (Villa and Houghton 2005).

In 1963, Texaco Inc. (then Texas Fuel Company) constructed the first oil wells in what would become the municipality of Orito. Between 1963 and 1976, an "oil fever" led to the construction of the Transandino Pipeline between Orito and the port of Tumaco on the Pacific coast as well as fragmented stretches of highway that increased the deforestation of the selva and propelled the establishment of the majority of urban settlements in Bajo Putumayo. While a small number of state-directed colonization efforts occurred in the mid-1960s and 1970s, the largest wave of contemporary settlement was driven by the expansion of commercial coca crops, which arrived in the region in 1978. The US-declared war on drugs reduced coca production in Peru and Bolivia, producing a "balloon effect" that led Colombia to become the primary producer of coca leaves in the 1990s while also maintaining its former role in the processing and international commercialization of cocaine. Coca cultivation in Putumayo continued to grow with the establishment of the 32nd Front of the FARC in 1984, and a few years later with the arrival of narcotraffickers from the Medellín and Cali Cartels. The Medellín Cartel placed Gonzalo Rodríguez Gacha in charge of operations in Putumayo along with his accompanying band of paramilitaries known as Los Combos and Los Masetos. The latter were expelled from Putumayo in 1991 after an intense FARC-EP attack and the civic action of the local population, who resisted paramilitary repression of civilians stigmatized as communists. That same year, the 48th Front of the FARC was born, and the FARC became the only leftist guerrilla group active in Putumayo after the M-19 (1980–82) and EPL (1983–early 1990s) guerrilla organizations abandoned the territory. Putumayo was granted state status under the 1991 constitution after being absorbed into different provincial administrative assemblages at least fifteen times over the previous one hundred years (CORPOAMAZONIA 2007).[15]

The community-led expulsion of paramilitaries may have contributed to their delayed return to the department, which occurred in 1997 when the AUC established the Frente Sur Putumayo, part of the Bloque Central Bolivar. The first signs of paramilitary social cleansing operations—selective assassinations, massacres, forced disappearances, and displacement of civilians purported to be guerrilla sympathizers—soon followed and continued until the AUC's official demobilization in 2006. Paramilitary occupation coincided with the channeling of Plan Colombia funds into Putumayo, intensified aerial fumigation, military surges, and the "securitization" of daily life that characterized then President Álvaro Uribe's national Democratic Security Policy.[16] An indisputable increase in violations of human rights and international humanitarian law by military and paramilitary forces was reported during the two terms of Uribe's presidency (2002–10) (MINGA 2008; Ramírez et al. 2010). Overwhelming evidence also suggests that mid-level paramilitary commanders failed to demobilize or simply reorganized into countrywide narcocriminal structures, such as Las Aguilas Negras, Los Rastrojos, Los Urabeños, Los Gaitanistas, and Los Constructores, which continue to operate in and beyond Putumayo. The Colombian government refers to these groups as *bacrim*

FIGURE 1.6 Small farm in the Andean-Amazonian foothills. Photograph by author.

(emergent criminal bands), or more recently, after the signing of the peace accords with the FARC-EP, as "post-demobilization armed groups" (GAPD) and dissidents. Paramilitary demobilization in Putumayo looked something like this: The state issued one hundred widows, a majority of whom were victims of paramilitary violence, twenty hectares of land for their collective sustenance. Fifteen minutes down the road, twenty demobilized paramilitaries received a hundred hectares of land to grow cacao as part of their official demobilization and civilian reinsertion process. I visited both farms after they were misted with glyphosate during aerial fumigation operations in 2007.

LEARNING "PROCESOS DE AMAZONIZACIÓN"

As we walked around the gardens, orchards, granary, solar panels, compost piles, and dry toilets during our visit to La Hojarasca, we listened to campesinos explain the workings of the farm school. They shared how it was a site for creative experimentation and articulation—*aprendizaje e intercambio* and *aprender haciendo* (learning and exchange, learn by doing)—between farmers, but also, of course, farmers and a constellation of organisms, elements, beings, and technologies. For some, it was a reencounter with diverse Amazonian seeds, trees, fruits, flowers, and plants. For others, it was the first time they learned to follow solar–lunar–nutrient cycles and to become attentive to a world of microbial metabolisms harnessing energy beneath their feet. La Hojarasca, and the other aspiring and working Amazonian farm schools I visited, are conceived of as places of *learning* in contrast to the model farms that form part of conventional state agricultural extension. Rural communities' frustration with the state's demonstrative workshop model was explained to me one day as *"No somos carros viejos para andar de taller en taller"* (We are not old cars to be rotated from body shop to body shop [workshop to workshop]). Instead of focusing on the transfer of knowledge and standardized technical models intended for replication from one farm to the next, the farm school methodology proposes to multiply agro-biodiversity across farms by engaging in the proliferation of what Heraldo calls *conocimiento vivo* (live or living knowledge). I return to the concepts and practices of living knowledge in later chapters of the book, but for now I mention that living knowledge forms part of experimental processes in gardens, orchards, and forests precisely because seeds, soils, plants, trees, and people come into being through recursive relationships. These relationships are always in the making, awaiting their next realization within particular socio-ecological conditions, acquired aptitudes, and remembered and expanding skills and imaginaries.

As I have begun to suggest, learning entails simultaneous processes of un- and relearning—on the one hand, innovating and recovering modes of inhabiting a selva world that is being degraded at multiple scales and temporalities by competing forces while, on the other hand, actively transforming individual and collective responses to this degradation. One must relearn how to relate to particular soils; nutrient cycles; rain patterns; hours of direct sunlight; animal, microbial, and insect behavior; vegetal life; and the interdependencies between watersheds, Andean foothills, and Amazonian plains. Some of the campesinos articulated this in terms of entering into transitional *"procesos de amazonización,"* which I translate as processes of becoming Amazonian human, or what Heraldo and MEROS refer to utilizing the categories that organize dominant gender relations in the region as becoming *hombres y mujeres amazónicos* (Amazonian men and women) (Vallejo 1993b). Much like Vinciane Despret and Michel Meuret's (2016) description of young people with urban backgrounds entering into processes of becoming shepherds and learning the practices of herding in southern France, learning selva in southern Colombia has to do with repairing ruptured relations and cultivating relations that have yet to become—that is, relations that one did not know. Put differently, procesos de amazonización are less about being and more about processually becoming with. This creative becoming begins with the revindication of saberes (know-how/wisdom) shaped by cultivating and being culti-vated by selva over countless generations. As I later expand upon, the affective resonances that result among farms and across a territory when "life makes life happier" is also a vitally contagious component.

Despret and Meuret describe the herding time that the creation of a flock produces in long-range transhumance as "composing with a place and a space in time" (32).[17] I conceive of the Amazonian human as one who learns to *compose and decompose into selva cycles*, following and hence entering into selva successional and ethico-relational momentums or the "flows of becoming" (Raffles 2002) that make a place.[18] In chapter 4, I engage with diverse agro-life practices that I conceptualize as *trajectories of selva apprenticeship*. Borrow-ing from Stengers and Pignarre, I argue that these "trajectories of appren-ticeship" (2011, 54) are invested not in the image of a consolidated movement of agroecological farms, but in the potential to multiply diverse procesos de amazonización in which the Amazonian human becomes one among many beings, organisms, and elements cultivating new ways of living, laboring, eat-ing, shitting, and decomposing together. I say beings, organisms, and elements strategically to acknowledge the presence of biologically and ecologically in-formed entities alongside other existents that do not fit into scientific or secu-lar categories. Integral farms, gardens, orchards, and chagras are composed of and by these heterogeneous assemblages.

What I would hear referred to as Amazonian, analogous, successional, bi-ological, and selva agricultures do not converge around a set technical model. Instead, they make an ethico-political move toward what Heraldo and others call a "technical emancipation of territory." The emancipation of territory—collective and individual, human and nonhuman to differing and ungeneraliz-able degrees—is from institutionalized forms of technical assistance, scientific expertise, and popular practices that do not support or emerge from living in and with a relational continuum of selva existence. For a growing number of the families I met, modernizing agricultural sciences are deeply embedded in capitalist structures that dispossess rural communities, turning them into consumers rather than producers of food and protectors or guardians of agro-biodiversity. These sciences rear their colonial heads when they are deemed "knowledge" that parasitically appropriates nonscientifically derived practices and/or dismisses them and renders them obsolete. This occurs when the latter cannot or do not aspire to demonstrate their scientific equivalence or become standardized models dictated by singularizing market logics and dominant intellectual property regimes.

For example, on several occasions, Heraldo demonstrated how, instead of sending a soil sample off to an urban laboratory and paying for its chemical analysis, rural families could compare the soil where they intend to cultivate with fecund animal manure on the farm. This is done, he showed me, by applying hydrogen peroxide to both the soil and the manure and comparing the intensity of the effervescent crackle of microbial life to determine whether a soil is healthy and apt for cultivation. The reason to avoid consulting a soil science laboratory is not just a question of reducing costs and external dependencies in a precarious rural economy where communities rarely have access to such technology. As I attempt to ethnographically demonstrate throughout this book, it emerges from the ontological differences between treating soils as artificial strata, or at best as a natural body that can be routinely chemically manipulated, and interacting with soils as living worlds that are inextricable from their ecological relationalities. As Heraldo engaged in the experiment, he told me that it was not a question of knowing, but rather of learning how to cultivate and recover various practices, aptitudes, dispositions, and affects. Heraldo, Nelso, and María Elva's emphasis on open-ended processes of learning that do not result in the accumulation of universally applicable knowledge reveals a tension that they and other campesinos maintain not only with the agricultural sciences and their ties to productivist imperatives, but also with the category of knowledge itself when it is separated from learning as a humbling and shared, as in multilateral, ongoing, and situated process. I am reminded of Achille Mbembe's description of Fanon's situated thinking as "metamorphic thought" (2017, 161–62), and the coconstitutive relationship that decolonial practitioner Paulo Freire (1970) proposes between knowing and learning.[19] Mbembe describes Fanon's situated thinking as "born of lived experience that was always in progress, unstable, and changing; an experience at the limits, full of risk" (161). Freire argues that learning cannot be understood as a mechanical transference of knowledge from the one teaching to the one learning, but rather as a process of constructing knowledge from knowledge that one already possesses from lived experience.

The tensions that campesino families manifest in relation to the universalist aspirations of the modern agricultural sciences is not because they entirely reject the teachings of soil science, ecology, or microbiology as rendered evident in the above anecdote about relating to chemical versus biological soils. Scientific practices that, on the one hand, support farmers' liberation from capitalist imperatives and extractive-based logics, and on the other, responsibly address Amazonian-based problems may be incorporated into their agricultural life projects. Simultaneously, these families engage with specific practices they learned from their parents and extended family members, and ones they continuously learn and recover in their exchanges with neighboring indigenous, Afro-Colombian, and other campesinos. This includes working with endogenous seeds and varieties of plants, trees, and animals that have been introduced into the region and demonstrate adaptive potential. For example, one day Heraldo shared with me how his Nasa indigenous neighbors taught him to plant in fields recently struck by lightning because these fields become more fertile. The Nasa had reached this conclusion by witnessing the upsurge of mushroom caps after a storm. Heraldo later read a scientific article explaining the way lightning bolts shatter nitrogen molecules, which then fertilize the air and penetrate the ground in falling raindrops. This was a case, he explained, where popular practices matched up with scientific ones. However, there are innumerable popular and ancestral practices that have no scientific equivalent and that form part of or are actively being incorporated back into farm, forest, and chagra life. Processes of becoming Amazonian human provide situated and concrete responses to a set of questions posed by feminist and Indigenous studies scholars regarding the decolonizing potentialities that may be unleashed when practitioners move "beyond simply selecting pieces of indigenous or alternative knowledges that appear to match with scientific knowledges" (Green 2013, 1). In these moments, contests over agro-environmental concerns may become able to acknowledge that different versions of "nature," different ways of knowing and enacting the world, are at stake.

ON THE SITUATED LIMITS OF ANALYTIC SYMMETRY

Within science studies scholarship there have been moves to "democratize" knowledge production in different global contexts under more pluralistic conceptualizations of science and modernity (see, for example, Harding 2008; Medina, da Costa Marques, and Holmes 2014; Rajão Raoni, Duque, and De 2014). More recently, scholars interested in decentering science studies from English and Euro-American analytics have proposed what they call a "postcolonial version of the principle of symmetry" to ask, "What might happen if STS were to make more systematic use of non-Western ideas?" (Law and Lin 2015, 2). Ethnographic conceptualizations at the interfaces of postcolonial and feminist science studies have made important contributions to our understanding of the kinds of ontological tensions that exist and are necessarily maintained between divergent knowledge systems and world-making practices (Cruickshank 2005; de la Cadena and Lien 2015; González 2001; Lyons 2014b; Verran 2001, 2002, 2013). Of course, beyond the confines of these academic debates, encounters between "Western" and "non-Western" ideas in the Americas have been ongoing since the Conquest and the control of the Atlantic after 1492. Focusing on the specificities of Spanish and Portuguese colonialisms, Latin American and diasporic scholars based in the United States proposed the "triad modernity/coloniality/decoloniality" as the analytical unit for understanding the way coloniality, the transatlantic slave trade, and processes of massive displacement in the Americas have been constitutive of modernity and the making of a capitalist world system (Castro-Gómez 2005; Escobar 2007; Giraldo 2016).[20]

In an epistemic sense, the production of modern scientific disciplines has occurred and continues to occur within asymmetrical power relations of ongoing coloniality. When indigenous, campesino, Afro-descendant, feminist, and popular sectors chant *"500 años de colonialismo"* (500 years of colonialism) during mobilizations across the hemisphere, they are engaged in struggles against specific forms of ongoing coloniality that are conceptualized in ways other than "postcolonial." This said, I am in no way arguing that postcolonial scholarship and subaltern studies have not been influential among political activists and scholars in and of Latin America. Nor am I claiming that a decolonial paradigm should become a singular explanatory tool to discuss the commitments and practices of diverse popular struggles and radical thinkers across the hemisphere (see also Pérez-Bustos 2017a). The historical production of scientific knowledge has always entailed its constitutive outsides in the making of the category of "science" in opposition to "religion," "superstition," "folklore," and "belief" alongside the ongoing appropriation and criminalization of nonscientific practices, and hence worlds, that has allowed techno-scientific practitioners to make claims to authoritatively "know" a singular reality. Examples relevant to this book occur in the encounters between agronomists and rural communities when the former claim to know what is and what is not a "good and productive soil," a "better breed of hen," or an "improved seed." Furthermore, as Helen Tilley (2011) and others remind us, the codification of "indigenous" and "traditional" knowledges as such is inseparable from colonial relations and imperial structures. In most cases, indigenous and "local" peoples have been forced to interact with colonial actors and structures, while the latter get to decide whether or not to engage with indigenous or popular know-how and "worldviews."

My intention is not to gloss over diverse scientific traditions by simply defining them as rooted in the projects and practices of colonialism. Nor do I underestimate the critical perspectives and subversive potential of scientists working within unequally distributed global positions. I am, however, interested in the limits of symmetry as a conceptual and political tool when placed in conversation with the kinds of alternative practices that Heraldo and the other alternative agrarian practitioners I came to know engage in when they attempt to transform their everyday relations, and strive to, as they call it, "decolonize their farms." There are important differences between the concept of symmetry as an analytical proposal and the ways symmetries and asymmetries are experienced, conceived, and enacted in the life proposals of Amazonian farmers. These farmers are not trapped in an either/or world that pits knowledge against belief. Nor do they make a multicultural or hybridist move to simply place scientific practices that perform "local appropriateness" in analytical and material symmetry with alternative or popular practices. El hombre amazónico is, in part, a name crafted by Heraldo to mark a contrast between the *"Amazónologo"* scientific experts who began to visit the region in the 1980s when what became known as the Amazon Basin was converted into an object of study and international environmental concern, and the local trajectories of amazonización of rural communities and alternative technicians who learn with la selva. As Heraldo explains it, hombres y mujeres amazónicos are not experts who come and go, sometimes to the benefit of local communities, but also often to their detriment, receiving public funds and academic credits to make technical recommendations for the agroecological systems of the territory. Even when they responsibly address Amazonian problems, scientific practices are categorically and not simply relatively different from the kinds of practices, and hence practitioners, that are cultivated when one lives, dies, and defends a territory under military duress.

Heraldo and others do not seek to democratize science—in other words, to open inclusive spaces for what are referred to as ancestral, traditional, and popular saberes within neoliberalized science policy culture, or to place science at the disposal of the interests of "civil society" as though a dualistic division exists between the two. The promises and practices of democratization may or may not take on relevance. They are always extremely situated political, social, and technical processes rather than universal aspirations. This is heightened in a situation where communities are criminalized due to their presumed engagement in illicit economic activities, their environmental activism and political mobilization, and the fact that they reside in territories that are also occupied and socially controlled by paralegal armed groups. By illicit economic activities, I refer to more than the cultivation of what have been denominated illicit crops in Colombia, Afghanistan, and other countries afflicted by war and narcotraffic. Neoliberal reforms have incrementally criminalized a whole variety of food production, commercialization, and seed-propagating practices. These reforms are progressively reshaping national agricultural economies and legislation in favor of multinational corporate chemical-seed-pharmaceutical conglomerates.

PLATE 1 Soil samples drying on the rooftop of the National Soil Science Laboratory. Bogotá, January 2009. Photograph by author.

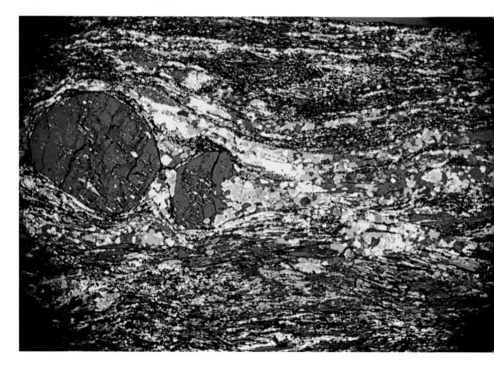

PLATE 2 Thin section of soil viewed under a petrographic microscope.

PLATE 3 Antonio's seed circle mística. San Lorenzo, Nariño, November 2010. Photograph by author.

PLATE 4 Wild passion fruit flowering in the creeping plant garden, March 2011. Photograph by author.

PLATE 5 Diverse Amazonian fruits and vegetables. Mocoa, Putumayo, January 2014. Photograph by author.

PLATE 6 Mural painted by youth in La Hormiga, Putumayo, depicting life before and after aerial fumigation with glyphosate, August 2007. Photograph by author.

Thus, while certain modern agricultural technologies are incorporated into rural communities' labor when they exhibit liberatory potential within the relational conditions of Andean-Amazonian ecologies, asymmetrical engagements between practices remain ethically and strategically important as a political or, better yet, life proposal. This is an asymmetry that subverts the authority granted to scientific knowledges and their nexus with capitalist accumulation over myriad other non- (or *not-only*) scientific practices and non/anti-capitalist ethics. By "not-only," I take inspiration from what Marisol de la Cadena conceptualizes as *excess* or "that which is performed past the limit" (2015a, 14–15)—in this particular scenario, that which performs past the conventional delimitations between "science" and "nonscience." For the campesinos I accompanied, the agricultural sciences must first demonstrate their alliance-building capacities with relational, more-than-capitalist worlds instead of obliging "local" practices to demonstrate their equivalence with the modern sciences. These kinds of asymmetrical analytical and material engagements struggle to resist the appropriation of popular practices by different scientific disciplines while acknowledging the ongoing debts these sciences owe to the same practices and practitioners they marginalize(d). This is not simply an inversion of symmetry. These rural communities are making a series of counter-dualistic moves that create openings for tense and potentially collaborative relations between scientific and "not-only" scientific practices. Beyond assuming fixed locations of subjugation that a "postcolonial symmetry" proposes to unravel, campesino families in the Colombian Amazon taught me the conceptual and political importance of considering decolonizing enactments of asymmetry. This book emerged from processes that displace the primacy of "knowing" in favor of ongoing processes of unlearning and relearning in such decolonizing efforts.

THINKING WITH LITTER LAYERS

Whenever I waited for Heraldo to hurriedly pack before we headed to the bus terminal in Mocoa to travel to a workshop, minga, or meeting, and (what became a running joke) inevitably forget seeds, farm designs, or the contact information for our future hosts, I perused the modest library on the open-air second floor of his farmhouse. The bookcases held agroecology texts by the Brazilian-based agroecologist Ana Primavesi, documents on different varieties of plants, animals, and the influence of solar and lunar cycles on tropical agriculture, essays on political economy and ecology, philosophical reflections written by Evo Morales, La Via Campesina, among others, and compilations of community development pamphlets. I was particularly interested in the things that Heraldo writes and designs: his blueprints for farms, orchards, and gardens, technical agricultural and ecological guides, and especially his manifestos and concepts of lo *Amazónico*.

On one occasion, I came across a series of articles by a veteran Colombian soil scientist, Abdón Cortés. Heraldo explained to me that he read Cortés's work because he finds him to be a "conciliatory" scientific ally for rural communities refusing to participate in extractive-based agriculture in the Amazon. Interestingly, Cortés, who is credited with institutionalizing the US Department of Agriculture (USDA) soil classification system in Colombia in the 1970s, would also be one of the first scientists to publish articles questioning its universal applicability—in particular, its taxonomic relevance for the country's tropical forests (Cortés and Ibarra 1981). Cortés managed the Subdirectorate of Agrology at the National Geographic Institute Agustín Codazzi (IGAC) for ten years. The IGAC is responsible for producing the country's official soil surveys, maps, and classifications and managing the National Soil Science Laboratory. Cortés was a leading scientist in the first modern national resource inventory of the country's Amazon, the Radar Geometric Project of the Amazon (PRORADAM), which was jointly financed by the governments of Colombia and the Netherlands between 1974 and 1979. He was also appointed dean of the former School of Agrology at the Jorge Tadeo Lozano University in Bogotá. I offer a small caveat to explain that I follow the terminology used by my scientific interlocutors, who refer to themselves as *agrólogos*—which I translate interchangeably as agrologist and soil scientist—even though the professional title does not have much current usage in the English-speaking world. I became intrigued with the idea of meeting this "conciliatory" scientific ally after I began to hear Heraldo repeat iterations of the following: "What we are saying, what we are doing, is thinking with Amazonian soils. It is from this creativity and a sense of territory that our life projects, a different life, can emerge."

When I initiated the fieldwork that would lead to the writing of this book, I imagined traveling to Putumayo to research the way stories about toxicity and illness in soils, plants, animals, and human bodies were able—or unable—to become evidence that travels and impacts political spheres within the context of US-Colombian aerial fumigation policy (Lyons 2018). However, in lieu of following the production of knowledge about nonhumans (i.e., glyphosate exposure in soils) and the agency that humans then attributed to them, Heraldo and others showed me that I would be better off allowing nonhumans to force, indeed oblige, me to think instead. Borrowing from Stengers, their proposal was akin to taking "seriously those nonhumans that are best characterized as

forcing thought rather than as products of thought" (2005b, 5) in the most reductive political sense as problems to be solved or situations that need correcting. Rather than focusing on how public policies directed from the sky dictate
the experiences and representation of life on the ground, I was encouraged to
think with the textured materialities, affective resonances, and recycling momentums making and unmaking the ground itself—that is, to follow the kinds
of potentialities striving toward existence as rural communities learn to relate
to organisms, beings, and decomposing elements that are *forcing thought* in
local Andean-Amazonian worlds.

I learned to situate La Hojarasca as farm school, resonating grasshopper,
and decomposing layers of leaves and rootlets within a host of indeterminate
and dynamic processes, such as flooding rivers, swamps, riparian forests, piles
of trash, manure, and porous membranes. William James (1996) refers to these
processes as "litter" (hojarasca) in the world. James reflects on those aspects
of being and becoming that many philosophical systems tend to ignore, try
to absorb into a transcendent purpose, or treat as merely transitional states
between solids and liquids, subjects and objects, stability and disequilibrium,
life and death, matter and form. He is wary of the philosophical drive toward
stabilizing the world as something definite, clean, and economically ordered.
In turn, Georges Bataille described *ambiguous matter* as the "unstable, fetid
and lukewarm substances where life ferments ignobly" (1993, 81). In conversation with the work of Caitlin DeSilvey (2006, 2017), what is of interest to me
is the potential for decay to reveal itself not only as erasure, but as a process
that can be generative of different kinds of knowledge, different forms of organizing, and different practices. DeSilvey is drawn to the storytelling capacities of entropy and decay in the discipline of critical heritage studies, which
tends to defend against rather than collaborate with evanescence. For me, the
metamorphic intermediacy of hojarasca calls into question fantasies of human mastery over, on the one hand, a world of solids and coherent states, and,
on the other hand, sociological preoccupation with the consolidation of mass
political movements—agroecological and otherwise. It was hojarasca that
inspired my ethnographic attention to material and immaterial processes of
composition and decomposition in specific scientific practices and in the cultivation of dispersed agro-life or selva agricultural processes.

Rural communities' growing attunement to the generative capacities of
these layers of decomposing leaves, stalks, pits, and fruit rinds convinced me
of the necessity to stay with the trouble—to borrow a proposition from Donna
Haraway (2016)—of the emergent and transitional as more than simply the
present captured in retrospect. For instance, the ongoing material risks and af-

fective tensions involved in transitional processes, such as leaving behind coca crops and learning to see the selva not as brush or weedy regrowth that needs to be cleared and tamed, but as food and remedy: the hopeful awkwardness of not knowing where one is standing and admitting it, of recognizing one's own participation in the environmental degradation of a territory, and attempting to understand and organize a farm as a biological corridor and forest reserve rather than only as a piece of property that is worked over and extracted from to obtain a family's economic sustenance. I was led to think carefully about the way life slowly grows in the midst of poison, and the temporal dynamics between moments of visibility and frenetic energy and periods of latency and imperceptibility in these processes. How does one write the times of glyphosate and the life span of such chemicals in the biochemical matrices of the soil? How does one attend to the gradualness and tentativeness of an economic reconversion when an open pasture returns to secondary forest?

Aguantar is the Colombian Spanish word that most closely translates to the English phrase "to endure." People often joke darkly that they are *expertos en aguantar* (experts in enduring). It can mean to hold on or hold out, to put up with, to withstand, and to bear. The popular expression *"no aguanta"* is an abbreviated way to say that something or someone is intolerable and should not be stomached. The form of life I attempt to articulate is a delicate balance between opposing, enduring, and transforming: in other words, opposing intolerable life conditions, carrying on within these conditions, and actualizing alternatives to the very conditions that make some lives and ecologies flourish at the expense of making others endure. The temporalities and materialities of hojarasca, its sinking in—decomposing into—the very thickness of a present filled with emergent potentialities and actualizations that may become frustrated, resurge, and enfold into one another again produces what I call a robust fragility (see chapter 4). This fragility supports tentative forms of socio-ecological repair and troubles commonsense, reductionist, and stigmatizing connotations of precarity and debility. Anna Tsing (2015, 179) proposes the word *resurgence* to articulate "the force of the life of a forest, its ability to spread its roots and runners to reclaim places that have been deforested."[21] The ecological relationality of the selva life and death I go on to talk about depends upon sinking into this deforestation, into the wounds, rotting materiality, and scars themselves.

2

THE THEATER OF LIFE IS ALSO
A STAGE OF DEATH

Beyond Surface Chauvinism

When I would look out across the rooftop of the National Soil Science Laboratory of IGAC, I would find myself viewing a mosaic of burnt reds, charcoals, tans, and coffee-colored hues. Tiny mounds of earth were laid out on scraps of brown packing paper. An equatorial sun beat down. Piles of dirt sweat. Oftentimes I would feel the urge to spin around fast enough until rows of soil samples became a pinwheel of shades of mustard, rust, the white of clouds, and sky blue. These samples had traveled from around the country to the laboratory in Bogotá in plastic bags from the soil surveys conducted by IGAC agrologists. It was during these trips to the rooftop to check on the progress of drying soils that I was inspired to ask: What times are we treading on? What ignored and multiple times? When everything visible to the naked eye was pushed up and turned over, sucked back down, and stirred around again. When the savanna surrounding Bogotá was undulating grassland now sealed over by expanding concrete and hardened like calloused skin after exposure to frigid winds.

Rural families living along the floodplains of the vast Andean-Amazonian rivers of Putumayo told me stories about the destructive *conejeros* (episodic floods) that inundate their farms. They place tiny markers in the streams to chart how fast the water is rising, on the lookout for electric eels and snakes that will be set loose on an unlucky dog, hen, or human neighbor. A thirsty ground percolates and then gurgles and gasps for air as it swallows up the garden and kitchen chairs. The air turns unusually cold followed by a haunting silence, an unearthly primal silence people say makes them want to scramble up the trees. This happens over and over again, every six years or so. These families count back as long as they can remember and take comfort in the tiny plants that will soon sprout out of seemingly lifeless debris. Fallen tree branches and fence posts rot and feed hundreds of imperceptible mouths. Butterflies and bees find their way back to the farm. Perhaps this cycle permits a kind of proximity to other times—times that interrupt a human life, inevitably press on, and double back again. These processes are not bound by progressivist, future-oriented temporalities or clear states between decay, disappearance, and regeneration. They might be conceived of as times beyond time that one has to enter into relations with in order to sense or imagine, much less to ever know. Michel Serres refers to this as the difference between living *in* time and living *out* in the weather (1995b).

There is something about how time is calendared within modern human lives that does not bode well for a life that is lived over thousands of years, or even one that can be destroyed after a few harvests. Who tastes when the soil is suffering? When is it possible to breathe in and recognize that it may be weary? In worlds produced by the ontological distinction between a living bios and presumed nonliving geos, soils often seem solid and ageless if they seem anything at all. Industrialized hands assume that they can exploit the same ground from one year to the next if by way of nothing else than the force of mechanization and chemical input substitution. When yields and revenues decline or fields dry up and blow away, there may be a flash of remembrance—the possibility of being obliged by the long hours of rocks in concert with the seasonal shifts in agricultural crop cycles and the velocities of microbial processes of decomposition and germination.

A soil scientist at IGAC once described to me what I repeatedly heard characterized by scientists as the soil's "troubling anonymity" in terms of seeing a tuft of hair, but not a face. He told me, "we see grass to be cut, plants to be harvested, and trees to be felled," or a vessel where life grows up and out but does not inextricably intermingle. Ecologist David Wolfe (2001) called this a "surface chauvinism," which fails to recognize that human existence is not separate from all that is out of sight and below one's feet. However, a microbiologist friend, Javier, whom I met when I was a research assistant at the Agrarian Microbiology Laboratory of the Biotechnology Institute of the National University (IBUN), may have gotten to the geological core of the matter.

He explained to me that it was hard to conceive of how to protect something that is not only incessantly trampled over, but that when it is imagined, it is imagined to be generated from rock—the assumed lesser important material of "nonlife." Elizabeth Povinelli (1995) instructed us that the attribution of agency, sentience, and intentionality of the sort that emerges with life (bios) from such materials as rocks is no small matter. It has long been the grounds on which to dehumanize colonized and enslaved peoples for their so-called premodern mentalities. How might this dualism separating a living bios from presumed nonliving geos be shifting in our postindustrial present when the impact of specific human trajectories on the planet has provoked a crisis whose effects are, as Donna Haraway says, "literally written into the rocks" (Haraway and Kenney 2015, 259)? Soils always already trouble this and other modern dualisms because they are processes of "geos" in transformation—rocks being worked over until they are teeming with and able to sustain life.

In this chapter, I am interested in the ethico-political implications that the shifting values of soils as living worlds have for the entwined fate of soils and state soil scientists. Ian Hacking astutely observed the incommensurability between the natural and physical sciences, or what he referred to as "a large number of only loosely overlapping little disciplines many of which in the course of time cannot even comprehend each other" (1983, 6). The transdisciplinary makeup of soil science and the pecking order that characterizes these heterogeneous disciplines are not exempt from a similar disunity. I learned that layers of ontological dissensus lay beneath the apparent unanimous agreement among scientists over the soil's marginality, ambivalent ecological significance, invisible labor, and failing health. What had gone wrong? Who was to blame? More importantly, how should policymakers, industrial trade associations, and scientific practitioners respond? These were some of the debates I heard during the 2009 Year of Soils campaign that was organized by IGAC with the support of what was formerly called the Ministry of Environment, Housing, and Territorial Development (currently named the Ministry of Environment and Sustainable Development).

The campaign had an overtly "geontological" orientation, borrowing a term from Povinelli (2016). It was deeply invested in bringing about the material and metaphoric visibility of a living soil, in stark contrast to a scientifically debunked yet popularly enduring notion of a quasi-nonliving physical-chemical container for the growth of land plants.[1] At the same time, soils complicate dominant biological definitions and understandings of life (bios) because, as I return to later, they do not possess the self-oriented reproductive capabilities attributed to most organisms, even if they are said to have a genesis and a life span. Soils are made of what scientists consider living and nonliving components and are the products of evolving relations rather than reproducible selves. I was intrigued and admittedly troubled by what I observed to be the dual political effects of the IGAC campaign, which rendered a specific biologically informed notion of liveliness visible while also maintaining certain silences and erasures regarding the constitutive role of violence in the conditions of existence and labor of the country's soils. In this chapter, I reflect on a natural resource campaign that urges people to place their feet on a receptive, albeit ailing ground. I go on to speculatively imagine what I call a poetics of "soil health" that takes inspiration from scientific conceptualizations and poetic forms of sensing without erasing the immanent role of violence in the lives and what, thinking with environmental movements in Colombia, I refer to as the "environmental" or "biocultural" memory of situated soils.[2] Thus, while Povinelli is concerned with charting an ontology beyond a binary of life and death, I make a slight shift to interrogate how death and decay are on a continuum with life.

OPERATIONAL ENCLOSURES

There is no single genealogy from which to trace the disciplinary consolidation of soil science. Demarcated scientifically and etymologically from the broader concepts of land and geologic strata by the late 1880s, the science of soil is considered to have been founded simultaneously in Russia and the United States. It was historically divided into two main branches—pedology, the study of soil genesis and classification, and edaphology, the study of soil fertility, use, and influence on living things—as well as four discrete subdisciplines: soil chemistry, soil mineralogy, soil physics, and soil biology (Churchman 2010). These four disciplines shape the National Soil Science Laboratory of IGAC. In Colombia, the first scientific observations of soil were recorded as early as the botanical expeditions of Alexander von Humboldt and Francisco José de Caldas in the early 1800s. After Humboldt crossed the Andes and arrived at the port of Guayaquil, he drew a cross section of the continent in which he displayed particular agricultural zones and labeled one of the columns at the side of the map as the "culture of the soil" (Güttler 2015). However, it was not until 1915 that the National Agrarian Institute was founded in Colombia and managed by Belgian scientists, marking the birth of agronomy in the country. Soil science textbooks in Colombia cite a slightly later period between the 1930s and '50s as the age of modern agriculture and the consolidation of a national soil science through the accumulation of state soil surveys. These surveys were used to establish a national cadastral to determine property value for state tax collection. By the mid-1950s, a generation of Colombian scientists had been trained in US and European universities. Some had engaged in graduate studies with the late Hans Jenny, an internationally renowned soil scientist at the University of California, Berkeley, while others worked with scientists from the US-based Rockefeller Foundation, which sent a mission to modernize Colombia's Ministry of Agriculture, similar to the foundation's postwar agricultural development interventions throughout Latin America (Shepherd 2005). In 1955, the Colombian Society of Soil Science (scss) was founded and began publishing the national journal, *Suelos Ecuatoriales.*

I would frequently hear IGAC agrologists lament how their discipline had become so firmly tethered to agricultural production, specifically the national coffee, flower, banana, palm oil, and sugarcane industries, at the expense of situating soils in their multitemporal and environmentally integral roles. This tethering to agriculture would leave a lasting asymmetry between chemists and physicists, on the one hand, and soil biologists and microbiologists on the other. The latter gained influence decades after their counterparts who deal with soil fertility and structure, both of which are prioritized within dominant agricultural paradigms as the properties of soil most directly tied to its productive capacities and chemical manipulation.[3] It is arguable that German organic chemist Justus von Liebig, who developed the mineral theory of the nutrition of plants and then went on to invent the first synthetic nitrogen-based fertilizer in the 1840s, established the field of agricultural chemistry that would come to dominate the scientific study of soils and set "agriculture on its industrial path" (Pollan 2006, 146).[4] Biological analysis, for example, is not yet included in standard state soil surveys. IGAC microbiologists showed me their consultation logs, which were filled with requests by private farmers for soil analysis of "pests" and "infections." This was evidence, they argued, of a persistent lack of interest on the part of most farmers in the potentially beneficent and symbiotic relations that could be established with soil biota and that occur in a given soil's microbial ecology. Besides the Subdirectorate of Agrology, IGAC's other administrative departments are Cadaster and Cartography, and some IGAC employees insinuated that the agrology division is still operating only because its soil surveys contribute to the processes of value extraction and the commodification of land.

The heterogeneous discipline of soil science followed a trajectory similar to many sciences that gradually complicated initial minimalist conceptions of their objects of study. Soil scientists moved away from early concepts of soil as an abiotic substrate and mere container for crop nutrients, toward their current definition as natural bodies and living systems akin to organisms, oceans, and stars. Concepts eventually derived from systems theories and ecology led soils to be conceived in their relations with vegetal, animal, and microbial life, designating them as living and open systems in themselves, and as bodies within larger open-ended systems of which they form part and also sustain, and through which they acquire their particular capabilities and limitations. Soils are definitionally composed of solids (minerals and organic matter), liquids, and gases that occur in a surface layer of the earth. Thus, they occupy a unique interfacial position between what the natural sciences have divided into atmosphere, lithosphere, hydrosphere, and biosphere. The inextricable synergy of mineral, water, air, and living matter emerges from the metamorphosis of solid rock (what is referred to in soil science as parent material) under the action of time, climate, topography, and living organisms, ranging from the activity of microorganisms to larger creatures such as worms and humans. Jenny (1941) developed the concept of the five "soil-forming factors" to describe the processes and actions that work upon and transform parent rock into living matter in his classic textbook, *Factors of Soil Formation*.

In the wake of the human agency and political and economic structures that went into creating the devastating Dust Bowl phenomenon of the 1930s in the United States, these formation factors came to be more explicitly understood as natural and cultural, even as human activities continue to be separated out because they tend to have greater impacts in a shorter amount of time than have other organisms.[5] Soils straddle modern nature/culture dualisms, but as I return to later, not symmetrically so. There is no such thing as a dead soil, for to *be soil* in scientific terms necessarily implies that not only is a wide range of biological activity sustained, but that organisms live within and make soil. However, conventional biological notions of life do not consider soils to be organism-like beings with self-organized bodies and the self-directed capacity to multiply on their own. Soil scientists explained this to me in conventional reproductive terms: "You do not find two masses of soil where there was previously one." Unlike chemical components, organic molecules, and organisms that come to be scientifically known through processes that render them discontinuous from one to the next, soils are described as a continuous skin stretching over the surface of the earth, interrupted only by bodies of water and rock outcrops. As such, they form an ever-changing terrestrial mosaic even when sharing similar parent rock material and climatic conditions. I listened to debates among scientists about whether soils are renewable, which, of course, depends upon the timescale being prioritized, and whether they are finite or reproducible through human intervention or bound by a set physical limit on earth, which is also a temporal question. Soils appeared to be both renewable and nonrenewable and simultaneously naturally finite and anthropogenically worked over, although not necessarily engineered.

Scientists have long dealt with the material and conceptual messiness of separating out "soil" from what they call, on the one hand, "nonsoil," and on the other, the larger terrestrial ecosystem. This difficulty is evidenced in canonical textbook questions, such as: How does one decide where the epidermal layer of a tree root ends and soils begin when they are always coevolving? How does one know if they are looking at a soil or a swamp after heavy rain (Jenny 1980)? It is the "involutionary momentum," borrowing a term from Carla Hustak and Natasha Myers (2012), of soil, rootlets, raindrops, and scientists that creates a complex relational arrangement that is ontologically challenging to delineate. From the scientists' standpoint, infinite potential exists for a new object to appear, continuity and discontinuity are blurred, and hence what they call "operational boundaries" are made to create the working fiction of a clear-cut world. Soils as temporarily observable objects of study and care are produced through these arbitrarily orchestrated enclosures even when they defy the summation of parts. Karen Barad refers to this as the enactment of "agential cuts" through which the components of phenomena become determinate within inherent ontological indeterminacy (2003, 815). I build upon this point in the following chapter.

At the soil science conferences and seminar events I attended during the Year of Soils campaign, members of the Association of Engineers explained that among their colleagues the soil is "everything that is not rock," while agronomists clarified that only the first one or two meters of earth's surface receive attention as soil in their discipline. The cuts made between bios and geos are profoundly politically and economically significant. They establish a given soil vocation and value to be extracted, as well as urban building codes, chemical input substitution plans, technological packages in agriculture, and which experts to call upon to "treat" the problems of the soil and to rally for their environmental conservation. At the Fifteenth Colombian Soil Science Conference in the city of Pereira, one soil scientist interrupted a discussion about how to build consensus around a definition of the soil, saying, "What we need is an operational soil that gathers together its primary essences. It would be like the 'theory of everything' proposed by physicists." Another scientist suggested establishing a "transdisciplinary agreement" that would coalesce around a coordinated concept that would be both flexible and robust enough to convince policymakers of the soil's strategic importance. Furthermore, this concept needed to be aesthetically performative in order to garner broader public appeal. The soil was often depicted as an "ostensibly boring element" that does not possess the inherent charisma of, for example, the country's pink river dolphins in the Amazon.

I listened to engineers, agronomists, and agrologists speculate about how to mobilize the soils enacted by their respective disciplinary practices, much in the way that science studies ethnographies have demonstrated the way different objects that go by the same name come to "hang together" through more or less coordinated processes of translation (Law 2004; Mol 2002). When the atmosphere was lighter, some scientists joked that if philosophers were unable to agree upon or ontologically stabilize their object of study, why should they be ashamed if the same was the case for their object of affection and inquiry. However, as IGAC's Year of Soils campaign revealed, scientific conceptions of soils as living systems had largely failed to successfully translate into policy mandates, environmental legislation, dominant economic models, or public imaginaries. The stakes of this failure, it was argued, were an alarming matter of concern for not only the quality and health of soils and the professional standing of their human specialists, but also for a range of planetary processes affecting the viability of all terrestrial life.

"PON TUS PIES SOBRE EL SUELO" (PLACE YOUR FEET ON THE SOIL)

The rapid sequence of PowerPoint slides was dizzying. With each flash of a map, statistic, and newspaper headline the fate of terrestrial life seemed to literally blow away in clouds of dust, be carried off by waterways as sediment, or risk being swallowed up by an eager and jealous sea. The launch of the 2009 Year of Soils campaign exuded an almost cinematic quality meant to shock audiences and to provoke a sense of urgency and moral accountability. Eighty percent of soils in the Andean region were reported to be suffering from erosion, and more distressing still, the same was said for 50 percent of Colombia's national territory. Colombia came in fourth behind Brazil, Argentina, and Chile in soil degradation in Latin America. A 2008 newspaper headline warned, "Erosion Eats the Caribbean Coast!" And between 1993 and 2003, the country's environmental authorities invested only 2 percent of their budget in *el recurso suelo* (soil resource), the term most readily used by IGAC agrologists. While real-time data were impossible to obtain, scientists at the Netherlands-based World Soil Information Foundation (ISRIC) estimated that 8.5 hectares of productive land were being lost per second throughout the world.[6] Given that it takes a thousand years to generate three centimeters of topsoil, the Food and Agriculture Organization of the United Nations (FAO) calculated that if the

FIGURE 2.1 Year of Soils campaign poster, "The Future Is in Our Hands."
Photograph by author.

world's soils continued to be exploited at current rates there might be roughly only sixty more years of topsoil left. The *Time* magazine article reporting on the issue was provocatively titled, "What if the World's Soils Run Out?"[7]

Julián Serna, then subdirector of agrology at IGAC, began every campaign event with the same appeal. "For the well-being of the present and future of humanity, the soils are a natural resource we should rescue from anonymity." He informed audiences that Colombia intended to set an international example with their "creole-led" campaign. The campaign converted 2009 into the National Year of Soils and led to the passing of a resolution declaring June

17th National Soil Day.[8] Indeed, seven years later, the FAO declared 2015 the International Year of Soils. Serna conjured up humble images of the soil as a sick patient in need of specialized medical care; an orphan without a benefactor in charge of its welfare; a dirt-caked beggar ignored on the side of a busy street; an entity endlessly stomped on and stepped over without anyone noticing where they were planting their feet. Various experts from the country's public institutions invariably drew a stark contrast between the soil, on the one hand, and water, forest, air, and biodiversity, on the other. The latter, they argued, were natural resources that had successfully garnered public attention, protective legislation, increased stewardship, and economic investment in the country's modern environmental political sphere. It was ironic, they pointed out, that national and international environmental authorities had spent the last fifteen years catapulting these other resources into public policy when the capacity of all terrestrial life to survive and flourish depends on the life-sustaining force of an unfortunately less charismatic partner—the soil.

I read the campaign as bringing from background to foreground an awareness that life occurs *in and with* the soil and not *on* it. "*Pon tus pies sobre el suelo*" (Place your feet on the soil) was the campaign's closing motto in December 2009. The slogan is suggestive on various material, figurative, and ethical levels: from a corporeal invitation to sink one's bare feet into earthy humus to a call for a renewed awareness that humans and other creatures live not stranded on a closed surface or solid ground but immersed in their mutual permeability and binding accountability with an open one. The relational underpinnings of the motto resonate, in part, with indigenous and environmental scholars Tuck and McKenzie (2015), anthropologist Tim Ingold (2011), and feminist geographer Doreen Massey (2005), who have problematized the idea that terrestrial life simply happens atop a benign surface. Serna made impassioned calls for scientists, policy makers, and ordinary citizens to reclaim the dismissed and overlooked "breathing skin of the earth."[9] "Soils have quality and health. They have life spans. They have vocations and untapped potential. They are our greatest natural capital. Can anyone say that we are exaggerating? Does anyone think that we are inventing a problem?" He repeatedly posed this rhetorical challenge to audiences, taking a dramatic pause as if waiting to see if anyone would dare to raise their voice to object.

Year of Soils events did not attract large crowds. In fact, no campesino, indigenous, or Afro-Colombian associations were invited. For many of these rural communities, relating to a living world beneath their feet would not have been a novel proposal, even as they continue to endure the colonial and modernizing effects of industrial ways of knowing and relating to what they may or may not conceive of as an object called soil. Furthermore, as I discuss in the next chapter, the kinds of lively existents and relational continuums that rural communities foreground may not be the same understandings of life that soil scientists seek to enact and render visible. Nor were social scientists, other than myself, or members of the general public present at campaign events. I was invited to attend after I made a special request to Serna, who decided that an ethnographic engagement with state soil science might somehow be beneficial to IGAC's goal of converting the soil into a matter of public concern. The organizers explained that the campaign was designed to strike at the heart of academic and governmental institutions and private industry, and then, as is common in state participatory discourse, be socialized at the regional level through these same institutional channels. Most of the yearlong debates circled around the political semiotics of representation, that is, how diverse "soil spokespersons" might conceptually conceive of and enact the soil in a way that would improve its legal standing, rational management, and health. A conceptual redefinition necessarily implied confronting the ontological complexities that were perceived to be obstacles for the "soil resource" and the efficacy and political relevance of its experts. The Year of Soils campaign attendees generally included functionaries from the Ministries of the Environment and Agriculture, IGAC, the Institute of Hydrology, Meteorology, and Environmental Studies, and the Association of Regional Autonomous Corporations and Sustainable Development; members of the National Cattle Ranching, African Palm, and Cane Sugar Trade Associations; representatives of the Cerrejón Coal Mine and the broader energy sector; faculty from the Agrarian Sciences Department of the National University; diverse scientists from the Associations of Agrologists and Engineers, the Colombian Society of Soil Science (SCCS); and employees of the National Soil Science Laboratory.

It was unsurprising that this set of actors struggled to maintain a tense balance between acknowledging what I came to think of as *soil as living system*, situating its ecological conditions of existence and rights to health for its own sake, and employing *soil as laborer*—the preoccupation with ensuring its economically productive capacities, future-oriented ecosystem services, and monetary value. From my work years earlier as a union organizer, a visual image of the soil as an exploited worker often came to mind when I listened to Colombian technocrats warn that the nation's soils would be called upon to endure intensified challenges. The impending signing of a free trade agreement with the United States, which was implemented in 2012, required the country to increase its industrial agricultural and livestock competitiveness, production of biofuels, and status as a biodiverse reserve for emerging international green markets. Beyond the national scale, the 2009 FAO and the UN Millennium Development Goals had enlisted soils to provide solutions to an array of pressing human predicaments, including global food security, extreme poverty, and expanding urbanization. Issues of *National Geographic* and *Nature Geoscience* highlighted the soil's provision of key ecosystem services, such as carbon sequestration, nutrient cycling and waste management, safeguarding water reserves, and providing a fit home for the planet's aboveground and enormous, mostly unknown below-ground biodiversity—speculatively referred to as "the next frontier of biotechnological advancement" (*Science*).[10] The list of current and future-oriented services that the soil as laborer was being called upon to provide kept on growing. At the Eighteenth Latin American Soil Science Conference in San José, Costa Rica, I witnessed one soil ecologist stand up and question his colleagues only half in jest, "What are we asking of the soil over the next forty years? That it resolves all of humanity's problems?" While his intervention provoked solemn nods, no one raised their voice to the contrary, even though the carrying capacity of "the world's soils" was projected to be at its tipping point.

As Benjamin Cohen examines in his book on the history of soil science in the early US farming context, what he calls the "soil identity" comes to be fundamentally understood as something improvable: "an entity governed by principles; as matter that could be studied, analyzed, and experimented upon; and thus, as something that could be made more productive" (2009, 129). The invention of agricultural chemistry facilitated a shift toward what I think of as the pharmaceutical treatment of soils as chemically manipulable bodies—bodies that must be kept healthy enough to endure intensified modes of work and whose stolen nutrients no longer have the possibility to be organically recycled and restored. Following the chemist Liebig and other analysts of the nineteenth-century soil crisis in Europe, in volumes 1 and 3 of *Capital*, Marx described the ecological contradiction between nature and capitalist society as "an irreparable rift in the interdependent process of social metabolism" (1991, 949). Indeed, "capitalist production," he argued, "only develops the techniques and the degree of combination of the social process of production by simultaneously undermining the original sources of all wealth—the soil and the worker" (1990, 638). It is the injection of capital into a particular soil, the so-called permanent improvements that change the physical characteristics, chemical properties, and biological life of that soil—that converts it into a laboring body deemed capable of being worked overtime, exhausted, and continuously resuscitated through chemical input substitution.

EVERYWHERE, YET NOWHERE

At the time of my fieldwork, there was no planning document for soils approved by the National Political Economic and Social Council (CONPES), even though CONPES documents existed for water, air, forests, and watersheds. This was no surprise, IGAC agrologists remarked, given that the soil was virtually excluded from the country's 1974 National Code of Renewable Natural Resources. More accurately, soils were not altogether absent, but instead occupied an ambivalent *betweenness*—intermittently enfolded into the regulation of other resources and the partial visions of sectoral interests without receiving independent or integral treatment.[11] Oscar, an agrologist contracted by IGAC whom I accompanied to conduct municipal soil surveys, which I describe in ethnographic detail in the next chapter, explained to me that policymakers could not determine how to treat such "promiscuity." Soils are everywhere and seemingly immutable within dominant human-oriented temporalities, too relational to be separated out—not in an absolutist sense, but in a disentangled enough sense—for the dismembering techniques of modern natural resource management: land is partitioned from water, subsoil from soil, flora from fauna, forest from watershed, and so on. Officials whom I interviewed at the Ministry of Environment claimed that the soil resisted this compartmentalizing "Western vision" to its own detriment and created a layering effect rather than an additive one through its transversality in the existing legislation. This kind of relational complexity resonates with other environmental situations that, as ethnographers have shown, defy bureaucratic categories that appeal to units and discontinuous concepts as well as to a commitment to methodological individualism through which experts are expected to know the components of a given situation and afford them agency (Fortun 2001; Murphy 2006; Petryna 2002).

My soil microbiologist friend at the IBUN, Javier, made a slightly different argument, focusing on the political stakes and contingencies of visibility. He claimed that it was precisely the soil's thick relationality that afforded it any, even if limited, environmental significance. He told me: "Soils have to nourish themselves from other laws. They remain unprotected when they are separated out. Simultaneously, these other natural resources depend on soils for their existence. If the watershed moves, for example, soils lose their environmental power, their ally. If they hired me to design a CONPES for the soil, *me quedaría enredado*" (I would get tangled up, or more figuratively, it would be an extremely convoluted task).[12] I spent six months accompanying Javier as a laboratory assistant during his doctoral research, which sought to identify the correlations between the presence of specific microbial consortiums and indicators of plant growth in rice. Javier's experiments kept us moving between sixteen different industrial rice fields, refrigerated bags of soil samples, a greenhouse lined with plastic cups full of this same soil and microbial inoculated rice seeds, and pages of statistical data. We focused our attention on the rhizosphere—a narrow region of soil that is directly influenced by root secretions and associated communities of soil microorganisms—what is otherwise known as the clingy clumps and granules of soil that do not fall away easily when one shakes a plant rootlet. Soil microbiologists consider the rhizosphere to be the most active region of the soil. The biochemical processes that make up this narrow milieu are extremely generative to think with—in particular, the complex microecologies that are created when clumps of soil do not detach from roots, microorganisms, and nutrients. All kinds of sensing, foraging, exuding, symbiosis, and parasitism make up this rhizospheric space. I learned that Javier was steeped in relational conundrums due to his methodological commitment to work with microbial consortiums as an alternative to adhering to a more conventional biotechnological approach. The latter would search to isolate individual organisms that could be heroically identified as responsible for the increased measurable growth in rice plants. Javier instead sought to trace the associative microorganisms of the root system in which no direct physical contact or attachment between plant and organism exists, but mutual benefits between them are realized. His dissertation committee members insisted that it was impossible to replicate the relational density of the diverse populations of the consortiums in the production of plant growth–promoting microbial inoculants that could be turned into standardized recipes for future commercial biofertilizers.[13]

During the Year of Soils campaign, some technocrats argued that it is the ontological ambiguity of soils that makes their legal protection so difficult. I heard one official ask, "How is it possible for soils to be everywhere yet to still be nowhere?" It was this question and the lessons that I learned with Javier and from the rhizosphere that struck me the most. Drawing upon Susan Leigh Star's (2006) approach to "residues" and "infrastructures," María Puig de la Bellacasa suggests that soil is an interesting case for the study of absence: in plain sight and also invisible. Approaching the soil as a "bioinfrastructure," (2014), she notes that the soil's passing into visibility reveals much about its ambivalent material and cultural significance when taken as an event in its

FIGURE 2.2 Inoculated rice seeds and soils in the greenhouse of the Agricultural Microbiology Laboratory of IBUN. Bogotá, November 2009. Photograph by author.

own right. If we focus less on the shift from "invisibility" to "visibility" and more on the soil's presence through absence, there is much to learn from the material and conceptual force of its ontological recalcitrance. In other words, what can we learn from the nonpartitioning of soil from world or soil from its thick and transversal ecological relations? More than "slipping through the cracks," an expression that I heard used throughout the Year of Soils campaign, soils might be better understood as *clinging to, and thus both filling up and exceeding the spaces between*—between disciplinary divides, environmental management categories, and scientific constructions of earthy bodies and boundaries. Soils work their way between the cracks, adhering to the soles of our shoes, where they discreetly travel with us long after we have trodden upon their backs, or in other instances leave muddy and exposed trails in our wake. The milieu of this middle ground is both troublingly messy and capacious. It raises the question posed by Star in her poem "The Net": *who will see the spaces between*? (1995, 31). Puig de la Bellacasa (2015a) reminds us that Star located in these fissures not only the violence and pain in the lives and labor of the erased, overlooked, and silenced, but also the spaces that are created by multiple and split identities—what I imagine as the ability to cling between, and thus to soil (as in to stain or muddy) dominant categories and habitual patterns of thought. Rather than rush toward an impulse for unobscuring clarity—that is, the separating out of soil from other resource categories and its thick relations—I am keen to ask: What can be learned from a state of existence that is as "clear as mud"?[14] This might be a way of sinking into the mud and composting into its complexities.

"GETTING OUR HEADS OUT OF THE SOIL"

What I found particularly interesting throughout the campaign and my field-work in different laboratories was the explicit shared nature of the construction of responsibility for the soil's perceived anonymity: subject and object had mutually produced and somehow failed each other. An ontologically complex object whose life span was temporally illegible in modern human-oriented timescales and sensory fields was said to be linked to a declining and uncreative subject whose scientific status was actively being depreciated. In most of my conversations with Colombian soil scientists, they described the socio-political treatment received by soils as a faithful reflection, indeed a mirroring effect, of their precarious professional position. The youngest soil scientists in the country at the time of my research were in their mid-forties. This was because the only academic department offering a professional degree in soil science was closed in 1993 due to reduced interest in agriculture, government funding cuts, and decreasing enrollment of students that began to effect soil science in many countries in the mid-1980s (Baveye et al. 2006). Between 1960 and 1993, the Agrology Department in the private university Jorge Tadeo Lozano in Bogotá trained more than three hundred agrologists in a "holistic agrarian and environmental vision of soil that is responsible for ensuring its integral management and conservation" (Malagón 2005, 79). I heard two different accounts of the closing of the Agrology Department. One ended with a dean screaming, "¡El suelo me sabe a mierda!" (The soil tastes like shit to me, or more euphemistically, I am over soils!). The other version cited a dispute between faculty members over whether the department should develop a more applied technical program or continue with an integral and exhaustive scientific one. Both stories also cited the department's lack of publicity, its failure to clearly demarcate agrology from agronomy and ecology, which were less expensive to study at the public universities, and the lack of profits generated by low enrollment in the major as reasons for its termination in a private university setting.

The Jorge Tadeo Lozano University was founded in 1954 with the aim of continuing the cultural and scientific work begun by the botanical expedition of the Nuevo Reino de Granada. The university first opened an Indoamerican Department of Natural Resources, but felt that the degree was too broad. Given the number of forestry engineers, agronomists, and biologists in the country, but the complete absence of specialists in soils, the Department of Agrology was opened in 1960 and produced its first graduating class in 1962. It was the only Department of Agrology in Latin America at the time. Given that the Tadeo is a private university, low enrollment in the major led it to be relegated to night courses and then eliminated altogether. Agrologists told me that when they have approached the public National University about the possibility of reopening the major, they had been told that the disciplines of agronomy and geography are sufficient to cover the study of soils.

Over the last twenty years, IGAC agrologists have witnessed the dismantling of their professional niche, which they argue has been picked over by an assortment of vulturine experts—forestry engineers, biologists, ecologists, and especially agronomists—who step in to speak on the soil's behalf and even fill positions specifically calling for agrologists. References were made to a historical antagonism between agrologists and agronomists that led many of the country's most seasoned agrologists to refrain from participating in the Colombian Society of Soil Science (SCCS). Agronomists were characterized as focusing on issues related to soil management, but within the agrarian sciences of the cultivation of land and crop production, whereas agrologists situate themselves as the *voceros* (spokespeople) and *médicos* (doctors) of the soil in its specificity and integral care.

In my interviews with Pedro Botero, a renowned soil scientist who had participated in extensive soil mapping of the country's Amazon over the last forty years, he told me, pointing to his full head, beard, and mustache of white hair and seventy-one-year-old body, "The fact that there are *viejitos* [little old men] who are still conducting soil surveys indicates an enormous lack of new soil experts in the country." Indeed, the scientific field of agrology has been largely masculinized in Colombia. According to the few female soil scientists I met, the levels of sexism that continued to permeate the countryside deterred women from pursuing the profession as well as the required amount of field travel, which was difficult to negotiate within heteronormative, patriarchal family structures. In 2017, the SCSS had 172 members: 125 men and 47 women.[15]

One female soil scientist told me that women still feel that they have to be especially "good" and "produce meticulous work" to be accepted in the field.

Several members of the Association of Agrologists jokingly confessed that they were too tired to fight for the soils, telling me, "*Estamos capando cementerio*" (which can be literally translated as "We are skipping the cemetery," or what I euphemistically translate as "We are cheating the graveyard" or "We are too old to still be alive"). Inspired by physicist Niels Bohr's popular and poetic statement that "a physicist is just an atom's way of looking at itself," I am provoked to ask what it means to say that a soil scientist is a soil's way of looking at itself.[16] Many of the country's agrologists inhabit aged bodies and have weathered faces. They are less agile but continue to physically traverse the country conducting state soil surveys. Some said that they felt stretched thin, conveying a sense of quiet erosion or leaching away of their corporeal and professional energies similar to the ways they described the soil's overexploitation. Despite the apparent shortage of soil specialists, younger agrologists, such as Oscar, who is a member of the last graduating class of agrologists at Tadeo University, told me that they were unable to envision a future with secure employment given that IGAC usually issues three- to six-month contracts. These short-term contracts for untenured employees were said to hinder project continuity and institutional memory in the Subdirectorate of Agrology. By the time I completed my fieldwork, Oscar no longer worked at IGAC.

When visiting the institute one morning, I overheard a conversation about how the agronomy and veterinary schools of the National University in Bogotá were reducing the number of hours that students had to spend studying soils.[17] When I traveled to the Biotechnology Institute of the International Center for Tropical Agriculture (CIAT) in the city of Palmira, my agronomist guide lamented that the soil division had been practically shut down a few years earlier. He explained that this was due to donor priorities shifting toward research on genetically modified seeds. Where these super seeds were to be planted, if not in soil, was a question he left suspended in midair. Not unlike other ethnographies that discuss how scientists in situated contexts are encouraged to get caught up in global projects and become conduits for dominant cultural and techno-scientific currents (Hayden 2003; Helmreich 2009), soil scientists throughout Latin America are also subject to powerful forces of diversification and change. These changes are perceived to be hitched to a planetary crisis of climate change, population growth, and the enduring consequences of the unbridled development of the chemical fertilizer industry and capitalist agricultural revolutions (Hartemink and McBratney 2008).

At the Eighteenth Latin American Soil Congress, a scientist from the World Soil Information Foundation (ISRIC) chided his colleagues: "We need to get our heads out of the soil and become politically relevant as more than doctors of dirt." Getting one's head out of the soil entailed scientists engaging in the double work of translating emerging global, political, economic, and environmental interests into the language of soil and translating the taxonomic technical language of soils into more malleable and user-friendly articulations and public imaginaries. In an aspirational sense, this would resemble the succession of displacements and changes in scale and alliance-building potential that Latour (1988) traced in Pasteur's ability to extend his laboratory "out" and bring the "outside" social world in. At the national level, processes of translation were embroiled in historic controversies between scientists over IGAC's institutionalization of the USDA soil taxonomic system in the 1970s. This institutionalization came at the expense of marginalizing more applied and indigenous classification systems and vernacular modes of identifying soils based on landmarks, topography, farm names, individuals, and affective relations. For example, Botero told me that he and other colleagues had been employing the FAO land classificatory system, and that they "tried very hard to do studies according to those who needed them." From his perspective, the FAO system considers the biophysical, socioeconomic, technical, and cultural conditions when determining the specific vocation of a soil rather than basing the vocation on the USDA's eight general classificatory categories that make broad statements about "apt" and "inapt" soil quality and use. I discuss the polemics surrounding the USDA taxonomy as it relates to Amazonian soils in further depth in the next chapter.

Botero also shared with me the mystical energetic forces of the universe that he feels are relayed through the soil; the sentient protective and demonic spirits that live in soils and ward off or summon snakes and land mines when scientists are in the field; and the primordial connectivity and intimacy that certain scientists say they feel when cradling dirt. These "mystical elements," he lamented, had been strangled out of the parameters of state soil science within which scientists are forced to work. He explained to me: "It's like my skin. My skin is a living being because it is part of me and I am a living being. The soil is the skin of the earth and because it is the skin of the earth it can be killed. This is a more comprehensive vision, but the idea that the soil has spirits goes even beyond this. I began to understand that we are part of everything through my work with the soil, my spiritual work."[18] Botero was the only soil scientist who spoke to me about spirits and cosmic energy forces.

Most of the agrologists whom I interviewed focused their attention on enduring disagreements over, on the one hand, whether they had alienated laypeople and policymakers with excessive technical jargon in addition to some of their colleagues who strongly adhere to other modes of classification, or, on the other hand, if laypeople were simply not "modern" enough, and that society at large should receive minimal training not only in soil taxonomy, but also in a generalized, what one agrologist calls *"alfabetización en suelos"* (soil literacy) (Burbano 2010).

Finding ways to articulate the problems of soil degradation to different publics is not particularly new for scientists. Degradation is depicted as temporally complex and multicausal physical, chemical, and biological processes of deterioration, such as erosion, contamination, desertification, and decline in fertility. IGAC agrologists' explanations for these phenomena ranged from the excessive use of agrichemicals and deforestation, to inappropriate technologies, poor planning and territorial ordinances, the imbalance between established soil vocation and actual use, or what they more generally referred to as *conflictos de uso* (conflicts of over- and underuse). According to IGAC, 29 percent of the national territory has use conflicts (Centro Nacional de Memoria Histórica 2016). One scientist crudely summarized the situation as: "All the problems of the soil are ultimately social problems"—the most cited example being that of the country's approximately 11.3 million agriculturally apt hectares of land, only 4 million hectares were being utilized for agriculture. At the same time, cattle ranching and grazing occupied 38 million hectares when only 8 million hectares have been technically deemed apt for these activities.[19]

Something that consistently drew my attention during the Year of Soils campaign was the almost total absence of any mention of gross income disparity, high rates of land concentration, historic lack of agrarian reform and the counter-reform implemented by right-wing paramilitaries, and more generally, the social, political, and economic impacts of over half a century of social and armed conflict in the diagnosis of the failing conditions of the country's soils and the underlying causes of degradation. As Diana Ojeda and María Camila González (2018) highlight in their work on campesinos and resource politics in the Colombian Caribbean, a recent study by Oxfam revealed that Colombia ranks second in Latin America in terms of unequal land distribution, with two-thirds of agricultural land concentrated in just 0.4 percent of farmland holdings.[20] According to official statistics, more than 7 million people have been forcibly displaced from their land (RUV 2016), and nearly 10 million hectares, or almost one-fourth of the country's officially agriculturally apt lands, have been appropriated by paramilitary groups (AMH 2009, 21–23).

The absence of rural communities and noninstitutionalized "soil" practitioners from campaign events went hand in hand with the relative exclusion of the violent trajectories and enduring dispossession that structurally shapes human–soil relations in much of the country, as well as the working conditions of specific soils as laboring bodies under intensified extractivism and military duress. Interestingly, five years later, in the context of the peace negotiations in Havana between the national government and the FARC-EP, the central organizing theme of the 2016 Colombian Soil Science Society conference was Healthy and Productive Soils for Peace in Colombia. This may not be particularly surprising given the current political juncture in the country and the way postconflict discourses are being taken up by all kinds of governmental and nongovernmental entities and international cooperation agendas. However, I am left with lingering questions about how to make peace with soils and what many indigenous peoples, Afro-Colombians, and campesinos do not refer to as an object called soil, but rather conceive of as a relation of which they form a part—in other words, how to make peace with soils when they were not considered actors, scenes, or casualties of war. As I go on to discuss, certain soil scientists have searched for literary methods in their attempts to articulate and render visible a specific quality of livingness in their object of study. It occurred to me to attempt an inverse experiment that I call the *poetics of the politics of soil health* to speculatively propose how the consideration of political economic structures, multilayered histories of violence, and the reconstruction of ecological memory might transform the analytics, sociopolitical relevance, and public outreach of soil science.

THE POETICS OF THE POLITICS OF SOIL HEALTH

Optical mineralogy is a gaze turned deeply earthward into seemingly dark, still, and silent depths. Indeed, when I first peered into a petrographic microscope at the National Soil Science Laboratory, I was slightly disappointed to find myself staring at what appeared to be an unassuming slice of magnified dirt.[21] However, as soon as the polarizing filter was slipped into place, uniform darkness exploded into a kaleidoscope of fuchsias, yellows, violets, and blues. Odd shapes took form, mutated, and then disappeared. Hues shifted in intensity from shades of light to dark, more radiant and increasingly dull as the light diffracted off mineral particles and the voids between them at different angles. The soil mineralogist who invited me to his workbench that morning registered my surprise and reminded me that this was only the color spectrum detectable to the human eye. He went on to measure minuscule quartz grains and the size of clay minerals, and to note plant fragments and channels that indicated good oxygen flow and porosity. For me, this moment was akin to what the well-known late US geographer and soil scientist Francis D. Hole described as the aesthetic "pleasures of soil watching" (1988). For the mineralogist, his optical measurements were important because they could alert him to early signs of degradation or other ongoing structural damage caused by climatic forces that are increasingly difficult to disentangle from histories of human use and abuse.

If geology tells planetary stories with the life cycles of rocks, and archaeology tells stories of human activity through the remains of material culture, soil mineralogy can be said to reflect the granular stories of human–soil cosubstantiation. I have reflected many times on my initial reaction when peering through the petrographic microscope—my impatient desire to see "something," and the illuminative sensations that accompanied the action of light piercing through somber shadow. I questioned why it was so hard to submerge myself and sink into the soil's darkness, rather than look for the diffracted patterns of minerals and imaginatively slip into the depths of the voids between. These voids have registered the vibrational motion of water molecules, films, and air deposited by wind and rainfall, and the compression produced by tractor tires and countless numbers of cattle hooves. They have inhaled within a body that has respired over and over again, often left exhausted and gasping for breath. Through the voids one can descend into the laggard times of weathering bedrock and the almost imperceptible leaching of one soil horizon into the next, only to extend with roots and sprout with tree seedlings as they strive to become a regrowth forest. The darkness harbors violent memories of being pierced open and sucked dry by oil wells, converted into property, demarcated by fences, and sealed under the asphalt of urban growth. When I focused back on the minerals emerging and transforming before my eyes, they became microbial remains rolled over by military tanks, displaced by synthetic fertilizers, and metabolized into colloidal slime that lines the tunnels of burrowing worms.

Soils defy modern dualisms between nature and culture, and "living" bios and the "nonliving" matter of geos. As such, they also trouble modern temporal divides between past, present, and future. There is no final material erasure of the past in the sedimented and residual fabrics of their recycling bodies. When a horrible event occurs in a place, many rural communities in Colombia and elsewhere say the soils, plants, trees, and other elements and beings retain this violence. Karen Barad alerts us that "the world holds the memory of all traces; or rather the world is its memory" (2010, 216). Co-laborer and host to all terrestrial experiments and tragedies, life sustainer, evolving body, grave, and trash dump, soils seem almost too resilient and hospitable for their own good. I am reminded of feminist philosopher Rosalyn Diprose's (2002) writing on the unequal distribution and selective forgetting of corporeal generosity.[22] Systematic forgetting results from the asymmetric evaluation of different bodies and the kinds of labor, benefits, and relations they enable. "Material" and Marxist feminist scholars, such as Silvia Federici (2018) and Alaimo and Hekman (2008), have paid close attention to the overlooked biological and reproductive processes that form the supposed passive grounds from which capitalist production occurs. Taking cues from these feminist analytical and political concerns, I ask: How might we work against the selective forgetting not only of the soil's labor and generosity—much of which is extracted by force—but also of its corporeal limits and relational conditions of existence?

In a 1984 interview, Hans Jenny said, "If you are used to thinking of soil as dirt, which is customary in our society, you are not keyed to find beauty in it" (158). Just plain dirt has often left soils appearing just plain dead, or the weathering rock that is exuding into the life-sustaining body of which my microbiologist friend Javier spoke. Both Jenny and Hole spent much of their careers inviting their scientific colleagues to turn their attention toward what I conceive of as material substance and its affective sensation, corporeal generosity and its poetic coalescence. They insisted that scientists needed to engage with artists and creative writers, if not become artists and poets themselves, if they aspired to transform the extractive and industrialized logics and practices that have come to characterize modern human–soil relations.

Over the last decade, the concept of "soil health" has emerged as an alternative among scientists, international agricultural development agencies, and resource-management communities to unsettle dominant perceptions of soil as simply an economic growth medium or container for crops. Health has largely replaced the expression "soil quality" that was common in the 1990s and that focused on individual traits within a functional group, as in the "quality of soil for corn production" or the "quality of soil for roadbed preparation," and so on.[23] A more integral shift toward health produces a certain friction with the notion of *la vocación única del suelo* (single soil vocation), which I often heard employed by agronomists and technocrats working at the Colombian Ministry of Agriculture. Single vocation refers to the soil's *singular* function—the labor and economic and environmental services it is deemed apt to provide. The acknowledgment that a soil has its own requirements and rights to integral care, well-being, and longevity is an important shift that foregrounds soils as living worlds with corporeal capacities and limits. Following Dimitris Papadopoulos (2014), it is possible to argue that notions of health reflect a burgeoning commitment to make room for alternative ontologies within techno-scientific practices, or to make time for soil as a relation of "care" beyond productionism (Puig de la Bellacasa 2015b). Yet, as Joe Dumit (2012) warns in his BioMarx experiment, health can also be an indefinite resource for market growth in which the health of the soil becomes an expanding capitalist arena for chemical-seed-commercial organic corporations and the imperatives of industry-funded science.

A commitment to work against the selective forgetting of the soil's corporeal generosity requires that health-based approaches be situated within the enduring violence and dispossession that binds the connective tissues, which may indeed be ruptured relations, between particular humans and the array of beings and elements that compose and decompose soils. These sedimented trajectories of violence—what, borrowing from Ann Stoler (2016), we might call "imperial debris" or "colonial presence[s]"—*are* the metamorphizing rock, decomposing litter layer, transplanted sediment, and soil granules clinging to forms of life that have been uprooted and subjected to extractive-based forms of death. Homogenizing claims of global soil degradation rest on geographically uneven conditions and are underlain by specific—often capitalist— notions of quality and health.[24] This raises a series of questions: How can soil health be assessed without first asking whose territories are being occupied; which enslaved and indentured bodies worked now-exhausted plantation fields; and which actors amassed tracts of land through violent eviction and illegal contracts? What are the political-economic processes that have converted specific neighborhoods into radiating toxic waste dumps while gentrifying and displacing residents from others? What kinds of residual debris are left in militarily occupied soils shelled by bombs and converted into mass graves and scorched earth? Furthermore, which corporations made these bombs and which armies and administrations gave the order for them to be dropped? What forms of life come to inhabit cratered soils, and which refuse to return to or cannot live in such a damaged place?[25]

When we look through the prisms of a petrographic microscope, we are not only observing the size and integrity of sand, silt, and clay particles. We are also ushered into dark shadows and permeable voids where the transformations of life, death, energy, and matter flow through and become hauntingly absorbed by a porous body. As an alternative to opposing or compartmentalizing techno-scientific modes of detecting soil health from poetic forms of soil sensing, what might be learned from allying both kinds of practices and their potentialities and limits?[26] What kinds of relations and sensing might the pairing of the sciences and arts aspire to restore by first asking: reparations for whom? At the very least, the dominant and reductionist logics of a singular soil vocation provision would not be left unexamined. At best, we might learn how to tell granular stories that work against the selective forgetting of the soil's corporeal generosity, a generosity that is unevenly distributed and always already composed of situated trajectories of human–soil ruptures and co-laborations.

EL TEATRO DE LA VIDA TAMBIÉN ES UN TEATRO DE OPERACIONES

[The Theater of Life Is Also a Battlefield]

During the course of my fieldwork, I met one scientist who had produced a poetic concept in his work to interest a broader swath of scientists and civil society in the life and vitality of soils. Veteran agrologist Abdón Cortés was one of the first soil scientists in Colombia to question the persistence of a mechanistic, as he called it, "three-dimensional treatment" of soils. He argued that this treatment was the product of Euclidean thinking, and that it had reduced understandings of soil to the still dominant physical-chemical (structure-fertility) paradigm at the expense of ignoring its biological components and ecological functions. His problematization of a mechanistic approach conjured for me William Connolly's (2010) discussion of "emergent causality" in that soils become in processes that are not simply an opposition between the mechanical, inorganic matter described by some physicists and the evolving self-organizing systems often described by biologists, but rather through processes that ontologically skew the distinction between organic and inorganic and sentient and nonsentient. Cortés proposed a five-dimensional vision of the soil that includes the empirical observations made by agrologists when they conduct soil surveys, the laboratory analysis of soil samples, and what he calls the "socio-economic and spatial-temporal dimensions" that constitute and are constituted by human–soil relations and the life cycle of a given soil (1991a). Many agrologists referred me to Cortés's now foundational texts that were based on ideas that began to take shape for him in the 1970s, long before modern environmentalist movements emerged in Colombia or the so-called ecological turn became prevalent in the wider field of soil science (Lavelle 2000). In one article, Cortés poetically refers to the soil as el teatro de la vida (the theater of life) (1991b). I have often reread this publication and his subsequent articulations of the soil as a "refuge for life" (Cortés 2004) to reflect upon the ways in which soils are more than a stage upon which human and terrestrial life plays out. The creative idea of a theater struck me in a more foreboding sense, however, when I heard military officials in Putumayo refer to their field of combat as a *teatro de operaciones* (battlefield or theater of military action).

Several years after our first encounter in Bogotá, Cortés explained to me in personal correspondence that his concept of the theater was initially inspired by the fact that "with a microscope, magnifying glass, or the bare eye it is possible to identify the existence of life in the soil at the most micro, meso, and macro levels": the lubricious movement of worms, the scavenging tunnels of beetles, the red bulge of legume nodules ingesting nitrogen from the atmosphere, and subterranean mycorrhizae cities bridging a forest.[27] These are but some of the dynamic processes through which soils ingest the life that treads upon and wiggles through their matrices, eventually curling up, disintegrating, and becoming metabolized, so that others may go on respiring within and sprouting from their earthy folds. This aspect of Cortés's theater resonates with soil ecologist Karl Ritz's (2014) filmic metaphor that presents a cast, set, and plot to introduce diverse soil biota and to discuss the biochemical processes at work in forming soils along with the multifaceted services they provide. However, beyond edaphic biota, Cortés's theater refers to a complex gamut of socioeconomic relations, such as soil–human health, soil–inequality, soil–illiteracy, soil–land occupancy, and soil–power, that situates political economic processes in a spatial-temporal and historical manner. There are brief moments in his writings when he suggests that the theater of life can quickly convert into a scenario of death.[28] I read the death of which he speaks as being one that is as much "naturally" catastrophic as it is "culturally" structural—natural-cultural through and through—from erosion-produced landslides and powerful floods that cannot be isolated from questions of how people end up living in precarious housing and on geologically unstable soils and floodplains; to the degradation that results from unjust land distribution and the historic marginalization and impoverishment of indigenous, Afro-descendant, and campesino communities; to the corruptive loopholes and mismanagement that undermine technical planning for soils, which are deeply embedded in ongoing feudal political, religious, narcotrafficking, and traditional landowning power structures (2004, 25).[29]

Step sound, step sound, step down with care,
take the shortcut if you dare,
leave the beaten path to piss,
pick fruit, seek shade, I hit or miss.

Step right, step left, step light my friend,
one step down to rise again
metal click beneath a boot,
a shredded leaf, a splintered root . . .

Ingold and Vergunst (2008) point out that while many ethnographers tend to carry out much of their fieldwork on foot, most tend not to reflect much on walking itself, but alternately on the destinations toward which people are bound, the multiple sites visited, or the conversations that happen en route. The same, of course, cannot equally be said of the practitioners and interlocutors that ethnographers tend to accompany. Rural communities in the Andean-Amazonian foothills are familiar with improvisational modes of walking over textured and shifting surfaces as well as the inevitability of navigating dodgy footing. For example, the deep sucking power of quagmire soils easily traps the unexpected hooves of cows and even swallows up entire baby calves. I remember the day a campesino taught me how to identify pockets of bog by tossing a rubber boot into what looked like an ordinary section of a field and then watching as the boot was noisily gulped down in a wet belch. Strips of wood are often placed in the muddiest pathways of a pasture. They bob between water and sludge, submerging and resurfacing depending on the patterns of the tiny waves produced by the wind and the weight of the human feet that seek bridging support. A hungry earth is said to crack open its cavernous mouth when a Christian god or other powerful forces are angered. One morning, when Heraldo told me that we would not be hiking around farms that day, I foolishly wore sneakers rather than my usual campesino boots. When our plans changed and we found ourselves in the midst of the selva, fire ants quickly scurried up my ankles, I went skidding down the first oily surface of a clay-laden gradient, and the only plant available to grab hold of to slow my descent was a pineapple that promptly stabbed me in the palm before I uprooted it with the force of my body weight.

Despite the delicate footwork required to negotiate the minor hazards, trips, and slips that threaten to lay them low, and the more consequential spiritual imbalances of an agentive ground, walking in territories that have been the epicenters of war or what were referred to as the country's "red zones" posed particular challenges for visiting scientists and local communities—human and nonhuman. In 2010, the pastoral social office of the Catholic Church in Putumayo organized a preventive campaign to encourage rural communities to be wary of where they were planting their feet. It was not safe to trust in the very act of stepping down in the first place. *"Pise seguro, salve su vida"* (step safely, save your life) was the slogan of a popular educational campaign on the dangers of antipersonnel mines. At the time of my fieldwork, land mines had been planted in thirty-one of the country's thirty-two states, and after Afghanistan, Colombia was registered as the second country most affected by antipersonnel mines in the world.[30] "Step safely" seemed to directly contrast with the IGAC campaign motto "Place your feet on the soil." The soil was understood not only as a generous and physically undergirding ally, but also as harboring an aggressive materiality that could potentially and violently disrupt the rhythms of everyday life and bodily integrity (Clark 2011). One most definitely could not and should not step down just anywhere. The prevention campaign warned people of being too curious, gathering fruit, standing too close to military barricades, straying from well-worn footpaths, seeking shade, and traversing remote corners of farms or pastures where armed actors may have accidently dropped or purposively sown military or handcrafted land mines during their patrols and irregular transits.

Almost inevitably, my conversations with scientists about Amazonian soils would turn to the well-researched and popular topic of *terra preta*, or "Amazonian or Indian dark earth." These soils have been of great interest to archaeologists and ethnopedologists for their anthropogenic evidence of millennia-old human activity in the Amazon, and more recently, to soil scientists and industrial sectors for their potential to be replicated as a charcoal-rich commercial amendment called biochar. While agrologists assumed that an anthropologist would be interested in dark earth, it was more infrequent to hear them mention the lesser known *suelos saladeros* (salty soils) as they are commonly called in Putumayo. Due to different factors, such as erosion by wind, the toppling of a tree, or the diminishing of water levels in rivers, different layers of soil have been left exposed without vegetation and with an elevated concentration of mineral salts. Unique meter-high, wall-like structures or more shallow pools of mud and water accumulate and attract all kinds of fauna that come to bathe, peck, gnaw on, and drink from the nutrient-rich clay. This includes chattering parrots, macaws, tapirs, monkeys, deer, jaguars, and humans who come to witness and hunt these noisy "messmates"—what Donna Haraway (2008) calls creatures that share a table, eat together, and from each other with all the ingestion and subsequent indigestion that occurs among living consortia. When I inquired about the whereabouts of the nearest saladero, some friends in Mocoa told me that they were less than a half an hour's bus ride away. However, they warned me that it was not wise to seek these soils out as they had been riddled with land mines. It was just carcass and feathers, they said. Four legs, two legs, what used to be legs, hooves, paws, and claws blown back and pointing high toward the sky. It was this image of a saladero cemetery that remained present in my mind when I listened to IGAC soil scientists chronicle their expeditions to the Amazon.

3

PARTIAL ALLIANCES
AMONG MINOR PRACTICES

*The "Elusive" Nature of Colombia's
Amazonian Plains*

After I engaged in several interviews with Abdón Cortés, he invited me to IGAC to share aspects of my research with rural communities in Putumayo precisely because agrologists do not consult with local communities about their economic visions for their territories or their everyday agricultural practices during official soil survey trips. However, I quickly found that the IGAC agrologists were more eager to tell tales of their early days of fieldwork in the country's Amazon Basin. Their stories always began with a surprising confrontation: what they described as the deceptive contrast between aerial views of exuberant tropical forest canopy and the less than robust and fertile topsoil they discovered lay below. The "enigma" of Colombia's Amazonian plains, as I sometimes heard them refer to it, mirrors long-standing tensions in neighboring Brazil and Peru that emerged between state planning for industrial agriculture and the reality of a soil deemed to be a serious obstacle to the development of a conventionally productive agricultural frontier (Fearnside 1985; Schmink and Wood 1992). This enigma partially harks back to what Raffles and WinklerPrins outline as a racializing genealogy bridging nineteenth-century anthropological theories of perceived agricultural backwardness in the Amazon—due to the supposed "indolence-inducing effect on race of a too fertile nature"—with cultural-ecological narratives of the late twentieth century, which describe identical social effects under a different but equally deterministic environment—a "harsh setting of nutrient-poor soils" (2003, 167–68).[1]

Soil scientists enlisted by the state in the 1970s to conduct the first modern national resource inventory of the Colombian Amazon, the Proyecto Radargramétrico del Amazonas (Radargrammetric Project of the Amazon; PRORADAM), were trained to work in the country's temperate interior and coastal zones where soils are generally a meter deep and acquire 90 percent of their nutrients from weathered minerals stored in their uppermost horizons. In stark contrast, the thin five- to ten-centimeter arable layer making up much of the Amazonian plains, where nutrient provision depends on a soil's organic phase, appeared to them more "litter layer" than "soil." In the accompanying photo (see figure 3.1), it is possible to viscerally detect the intimate and recycling interplay between soils and selva that makes for an ongoing relation rather than a stabilized entity. As I go on to describe, soils can exist only if selva, plants, and their corresponding microbial communities exist because the continuous momentum and temporalities of nutrient cycling mutually sustain selva, soil, and everything else below, above, and in between. It was rural communities growing attunement to the dynamic conditions of existence of this "litter layer" that inspired the name of the Amazonian farm school in San Miguel, Putumayo: La Hojarasca.

Agrologists, on the other hand, have produced a long list of naturally limiting factors to characterize the litter layer they discovered in the Amazon, including bad genetics, old age, mineral-deficient parent rock, acidity, and the tendency to erode rapidly under heavy rainfall once forest canopy is felled (Cortés and Ibarra 1981; León 1999). Furthermore, the dominant clay base of many Amazonian soils is an extremely weathered kaolinite with high levels of aluminum and iron oxide, rendering them toxically inhospitable to many conventional commercial crops without substantial "corrective" measures (i.e., lime and heavy fertilization). Cartoon depictions in one of the Year of Soils publications, *Suelos para Niños* (Soils for Children), introduce children to the "senile" oxisols and ultisols or red clay soils of the Amazon. *Ultisol* is derived from the word *ultimate* because ultisols are seen as the ultimate product of continuous weathering of minerals in a humid, tropical climate without new soil formation via glaciation or sedimentation across longue durée geologic time scales. As I discuss later in this chapter, these are only two of the twelve soil orders making up the US Department of Agriculture's (USDA) soil tax-

FIGURE 3.1 Layer of hojarasca in Putumayo. Photograph by author.

onomy. The USDA system was brought to Colombia and institutionalized in the 1970s by soil scientists who completed their doctoral studies in the United States. Of course, much archaeological research on the well-known *terra preta do índio* (indigenous dark earth soils) and rotational slash-and-burn agricultural techniques has problematized the racist idea that "impoverished" soils inhibited cultural development in a pre-Columbian Amazon.[2] I am less interested in these debates and more concerned with the particular conclusions and policies that resulted from what Fernando Franco (2006) calls the "scientific colonization" of the Colombian Amazon during the PRORADAM project (1974–79)—in particular, the way historically racializing scientific narratives

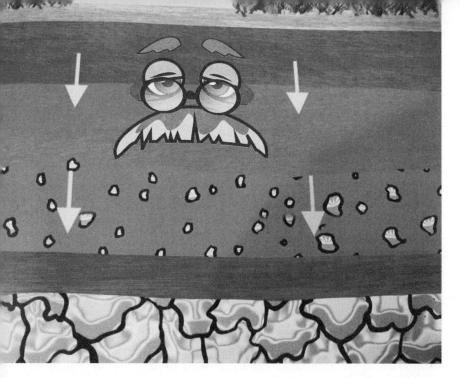

FIGURE 3.2 Depiction of "senile" Amazonian soils in *Suelos para Niños* (IGAC 2008, 107). Photograph by author.

about the social effects of "poor soils" became linked to the contemporary criminalizing discourses that underpin militarized US-Colombia counternarcotics strategies.

"Right off the bat, we say that a farmer is planting in degraded soil. This is a different soil. It is almost as if the organic material rejects mixing with the mineral elements. It is totally irregular," Abdón Cortés explains to me. We are seated in the Subdirectorate of Agrology at IGAC, where Cortés now works as a semiretired consultant. He goes on to explain that unlike conventional agriculture-driven colonization, where campesinos are not simply encouraged, but also legally required, to clear forest in order to work "better" and hence gain property rights over the land, this same practice in most of the Amazon yields only two or three consecutive harvests before soils are said to be *acabados* (ruined, washed up). This coincides with stories of the cyclical clearing of forest, exhaustion of soils, and eventual conversion of agricultural plots into pasture land that were told to me by many of the campesinos I met in and around Putumayo. The notion of "intentional labor" buttressing the titling of property in which a human, primarily masculine and able-bodied, subject is established through his capacity to "add to" and "transform" the land by intentionally harnessing an extraction process under subjective control, has long underpinned colonial and subsequent national political economic paradigms (Povinelli 1995). Clearing forest as a mode of establishing ownership is a state-supported practice ubiquitous throughout Latin America and many other parts of the world, precisely because states have formally and informally constructed forested landscapes as empty.[3] Above Cortés's head hangs a poster of dark, robust dirt cupped between two hands. "Care for our soils," it reads. "The future is not only in your hands, but also below your feet."

"Can you guess what our problem is right now?" he asks me in a booming voice that seems to ricochet off the walls of the small office where we are seated, drawing my attention back to a stack of documents that made up what at the time was IGAC's unpublished Putumayo soil survey laid out on the desk in front of us. "It is the same problem that we have had for the last thirty years. What are these soils good for? What do we tell the nation? Pasture lands, forest conservation?" He referred to a dilemma that had become as much a political problem as a scientific and technical one. "If we repeat what we said in 1979. . . ." Cortés's voice tapers off, and he remains tensely silent. One of his colleagues quickly chimes in, "Well, it would just be embarrassing." When I returned to IGAC almost a year and a half later to inquire about the status of the Putumayo soil study, I was informed that it remained "on stand-by" for further technical consideration. As I listened to these soil scientists, I tried to imagine Heraldo Vallejo crouched down harvesting root vegetables beneath a knotty mass of creeping plants while he explained to them that soil as a stable and quasi-independent entity would be hard to find on his farm. Indeed, there is no such entity that can be abstracted from an entanglement of ongoing relations that cannot be otherwise. While Heraldo and these scientists would likely agree that they were not in the presence of a "mineral-based soil," for Heraldo, this does not pose a problem that needs to be solved. Furthermore, the Amazonian agro-life processes that he and the other rural families and collectives I met are striving to cultivate depends upon *afinar los sentidos* (fine-tuning the senses) to what is going on in and all around one's body rather than looking down to diagnose and better manage an "irregular" or "poor" soil.

In this chapter, I place science studies analysis in conversation with farmers' practices to discuss the partially coinciding, diverging, and incommensurable relations that emerge between caring for the soil for the purpose of scientific interests and economic imperatives, and caring with a world full of beings that mutually nourish each other. At the confluence of war, repressive antinarcotics policies, and military-led development interventions, I begin to discuss the limits and possibilities—ethico-ecological imaginaries, economic pluralism, and material transformations—that emerge along with these different forms of relating, both for the lives of soils and for those who may or may not engage with "soils" as partners in and for life. The growing interest in multiple, nonscientific knowledges and practices at the interfaces of political ecology and science studies has provided pointed critiques of the politics of processes claiming to bring scientific and nonscientific (or not-only scientific) knowledges "together" (see Delgado and Rodríguez-Giralt 2014; Goldman, Nadasy, and Turner 2011; Heller 2007; Mathews 2011; Nadasy 2003; Tsing 2011). I am interested in the ethnographic moments when state soil scientists make fleeting moves to question their classification systems and the dominant productivist logic underpinning institutionalized taxonomy—specifically, when they attempt to respond to the agroecological particularities that they encounter in the country's Amazon.

I consider the moves of these agrologists as attempts at "becoming minor," along the lines of Deleuze and Guattari's (1987) proposal in which they draw out important differences and tensions between what they call "state" or "royal science" and "minor science." The latter, they argue, is more of an experimental practice that involves confronting problems themselves instead of theorems, looking for flux in both what is known and what is problematized, and resisting reproduction while following matter's immanent traits. Dimitris Papadopoulos (2010) refers to this description of minor science as a "surrendering to matter" (77) rather than the production of a science of matter or a technology to harness and control it. Furthermore, as Matthew Wolf-Meyer (2017) emphasizes in his work with Lacanian psychoanalysts, minor sciences rarely have any interest in attaining a dominant position but are more oriented toward their own perpetuation and keeping the science alive even if it remains marginal. Attempts at "becoming minor" are inextricably tied to scientists' struggles against the destruction of what allows them to think, imagine, and work in the midst of their institutional entanglements with state and capital. I became interested in the extent to which IGAC agrologists' attempts at producing a "minor soil science" potentiate alliance-building with the territorial visions of rural communities in the country's agricultural frontiers. While the latter may be perceived to be marginalized in a more traditional political-economic and social sense, many of the campesinos I encountered in Putumayo do not position themselves primarily as occupying a weak position or as disempowered. In fact, I found that it was scientists situated in the capital city of Bogotá who more frequently expressed varying degrees of constraint and marginalization depending on their proximity to and reliance on state and private research funding and industrial partnerships.

To discuss the ways Heraldo Vallejo and Abdón Cortés enact different, albeit interacting, entities when they say the word *soil*, I rely on what Eduardo Viveiros de Castro calls a process of "controlled equivocation" (2004). Uncontrolled equivocation refers to a type of communicative disjuncture in which the interlocutors are not talking about the same thing, and do not know this. However, these apparent misunderstandings do not occur because of different perspectives on a common world, but instead result when the interlocutors are unaware that different worlds are being enacted and assumed by each. In other words, "an equivocation is not just a 'failure to understand' but a failure to understand that understandings are necessarily not the same, and that they are not related to imaginary ways of 'seeing the world' but to the real worlds that are being seen" (11). Controlled equivocation would be the awareness of, or the making explicit that, a type of communicative disjuncture may take place when different realities or worlds meet. On this basis, a disagreement or struggle over the meaning of "soil" occurs because campesinos' and scientists' locally situated, albeit interacting, practices enact different worlds—worlds, by extension, of which an object called soil may or may not form a part. Along the same lines, the appearance of an agreement about the meaning of "soil" might actually occlude partial and radical differences among the scientists, bureaucrats, and diverse networks of farmers and rural collectives that I met and accompanied. The awareness that an equivocation is occurring is not a revelation available only to the anthropologist in her enterprise of translating or communicating cultural difference. The shifting value of soils for scientists and rural communities, and the respective ways that these values inform, are contested, or are rendered marginal by policymakers was a contentious topic of conversation among both groups, as I demonstrated in the previous chapter among IGAC agrologists. However, while Heraldo and other campesinos are able to control the equivocation when they speak of "soils," state soil scientists largely find themselves unable to do so. This is, in part, because they confront an element that eludes the scientific categories and practices they have produced to measure, describe, and employ it.

CAUGHT BETWEEN "LITTER LAYERS" AND AN EXPANDING SEA OF COCA

Commissioned by IGAC, the Ministry of Defense, and the Inter-American Center for Photointerpretation, and with Dutch funding support, PRORADAM combined remote sensory imagery with field studies to collect soil, vegetal, and other samples. The project categorized the country's Amazon into three large soil groupings from 80 percent denuded plains and 20 percent, formed from rock structures or sediment from Andean and Amazonian-born rivers.

It concluded that 1 percent of these soils are apt for intensive perennial crops, 18.3 percent for conventional agriculture and livestock, and 81.6 percent are unfit for agriculture and have severely restricted vocations (PRORADAM 1979). Furthermore, the final report assured that indigenous peoples throughout the region were the only "natural conservationists," leading campesinos inhabiting the country's agrarian frontier to be denigratingly represented as environmental predators by the political elites and academics who would influence the formulation of public policy post-1980 (Del Cairo, Montenegro-Perini, and Vélez 2014).[4] The early racialization of Amazonian indigenous peoples as developmentally backward and the later heroization of settler populations as brazenly expanding the agricultural frontier was inverted and also remained unchanging, depending on shifting political contexts and state economic and environmental priorities. In no instance in PRORADAM publications is the existence of indigenous or campesino soil knowledges and categories considered.

Just as the PRORADAM project was concluding, the intensified presence of illicit coca crops and their violent ties to paralegal armed groups in the Amazon captured not only the state's attention, but also geopolitical scrutiny from the United States. Contrary to agrologists' expectations, what would soon become a full-fledged war on drugs relied on military interventions to secure both rule of law and licit-based capitalist development. "What happened? Coca happened. Right when we published our findings the western Amazon became an obstacle for state security. We were unable to influence the region's economic development or even its conservation," Cortés lamented. Historically racializing discourses that purported to diagnose the social effects of "poor soils" quickly converted into criminalizing ones. When reading US congressional reports and USAID documents that attempt to analyze the failure of illicit crop substitution projects, I found that these reports began to suggest that the "impoverished" quality of local soils in the Amazon exhibited an inherent propensity for illegal economic activities and livelihoods.[5] The stigmatization of regional soils buttresses the criminalization not only of illicit plants but also of local peoples, and as I argue, the criminalization of entire ecosystems that continue to be subject to chemical warfare tactics through aerial fumigation and forced eradication policies.

Following Craib (2004) and Scott (2009), soil surveys can be understood as a specific technology used to consolidate the nation, one in which the scientific knowledge about soils generated through maps, field observations, and laboratory analysis produces both a classificatory device and the material infrastructure upon which national development intends to be built and carried out. It is no wonder then that Putumayo's soil studies have historically evolved in patchwork fashion and at lower resolutions. There were no consistent national interests attached to them until the advent of the US-Colombia war on drugs. It is soil as a classifiable entity with a potential working vocation that is of interest to state development imperatives, which compare territorial units on the basis of their productive capacities. Under public law, technical recommendations for soil use are intended to underpin territorial zoning plans that, in turn, inform municipal, state, and national development strategies (see, for example, Law 99 of 1993 and Law 388 of 1997).[6] Accordingly, more costly and detailed soil maps are generally commissioned by professional associations and private industries or through cofinanced agreements between private actors and the state. These studies are reserved for what soil scientists informally call the "more promising areas" of the country that possess the fertile and mineral-rich soils upon which capitalist agricultural development and other forms of economic growth depend.

Throughout the 1940s, soil mapping concentrated on the country's Andean center and gradually progressed outward. Pockets of information on the western Amazon exist for the more densely populated Andean-Amazonian foothills and early outposts of military colonization. Detailed studies of Putumayo's most fertile valley, Valle de Sibundoy, are also archived in the IGAC library. "We know what to do with these soils because they resemble the Andes. The problem is the Amazonian plains, and even more so, the plains impacted by coca, cattle, and agriculture in general. In addition to our uncertainty about their productive potential, good zoning is limited by the lack of public order," Marco, one of the agrologists who worked on the 2011 Putumayo soil survey, told me. He went on to recount how his technical team could only retrieve samples of soil from pastures or other deforested areas for fear of tripping land mines planted in the monte by armed groups. These scientists had to seek permission from the FARC-EP to enter their territorial strongholds, and this did not guarantee that IGAC's equipment would not be confiscated on the way out. After more than a decade of aerial fumigation and forced manual eradication of illicit crops, local communities are deeply suspicious of state employees, whom they perceive to be potential informers on the whereabouts of illicit coca plants. The list went on, and Marco imparted a scenario in which agrologists were caught between a rock and a hard place. From their perspective, this rock was weathering into a barely recognizable soil and the hard place was tropical forest besieged by a hardy shrub called coca.

IGAC agrologists, whose potential object of study was said to be elusive and difficult to access—thin, senile, violent, and dangerous—found themselves working for a state whose vision of the western Amazon has been monopolized by an itinerant sea of coca monitored in terms of hectares of land, and not an attentiveness to the particularities or alternative productive capacities of local soils. As perceived by the state and the knowledge it has been able to enlist, the classifiable entity that it intends to enroll in national development policies is either nonexistent or violently and illegally occupied. The president of the Colombian Society for Soil Science, a soil physicist and chemist from the city of Cali and the first female president in what I have mentioned is an overwhelmingly male-dominated profession, explained to me that over the last forty years the state had viewed the country's soils through a restrictive blinder: as a *medium* for the growth of illicit crops and a disputed *territorial component* occupied by armed groups that fractured and undermined the state's sovereignty. In either case, soil, as an object of study and a potential resource, is unable to emerge without networks of research that depend on shifting political conditions that extend far beyond the consent of the state. While I have begun this chapter by referring to a preexisting entity called soil, I go on to show how an entity called soil emerges—albeit partially or not at all—through the affective relations, embodied labor, and everyday practices of rural communities and state soil scientists. I first engage in an interlude reflecting on the extractivist objectives of the war on drugs and the socio-ecological underpinnings of territorial constructions of peace, which necessarily implies rethinking regional relations with soils-selvas.

WAR THAT GOES BY ANOTHER NAME

At first glance, it seems possible to situate the chemical warfare and overall eradication component of US-Colombia antidrug policy within other biopolitical histories and formations of killing: one in which a future peace is actively pursued through present acts of poisoning. It is a liberal way of making and justifying the ongoing nature of war that wraps various kinds of killing within an imaginary of safeguarding and life-making narratives. Killing is posed within a "future perfect" tense (Povinelli 2011a, 167) as a necessary redemptive mode of birth that will bring new economic beings and domains of social life into existence. Under the war on drugs this has referred to the geopolitical intervention of liberal moralities, such as rule of law, culture of legality, public health, and licit capitalist livelihoods. Eradication relies on a capacity to make live that rests on the necessity to make die, rendered evident in the dual meanings of the Colombian Spanish word *arrancar* (as in *arrancar la coca*, or as in how communities refer to manual eradicators as *arrancandores*). Arrancar means to start something—an engine, for example—and also to uproot or rip out, a violent cleansing or weeding that pries open the space for regrowth. A 2009 Volvamos a la Vida (Back to Life) campaign organized by USAID in Putumayo posed what one official called "reflexive questions about the social drawbacks of cultivating illicit crops." Postcards distributed during the campaign starkly contrasted black-and-white images with those in color. They read: "Live or die? Smile or cry? What kind of life are we sowing?" and posed a moral choice between pepper harvests or handcuffs, saxophone or headstones, goalie kicks or toes tagged in a morgue. Elsewhere I have discussed the way stigmatizing state campaigns framed as combating *la mata que mata* (the plant that kills) produce the cocriminalization of "natures" and "subjects"—plants, soils, and peoples. A criminalized nature is no longer the object of conservation or protection and is perceived to be in cahoots with a criminal subject that is rendered ineligible for humanitarian aid even when it is public policies that turn people into internally displaced populations and refugees (Lyons 2016a).[7] It is what results from this coconstituted stigmatization that I refer to as a *criminalized ecology*, which both perpetuates and is perpetuated by USAID's statements that "poor" soils have illegal propensities.

At stake at the heart of antidrug policy is not just the taking of the biological life of a plant, the severing of illicit human–plant relations, or the "correction" of criminal and acidic soils. Rather, it is the increasingly evident association between eradication efforts and the expansion of a national development model referred to as a *locomotora minero-energética*, which I roughly translate as "mining-energy locomotive" (Departamento Nacional de Planeación 2010). In 2011, more than half of Putumayo, along with several neighboring states, was reclassified from Amazonian territory to a Special Mining District, accelerating oil production from 8,000 barrels a day in 2000 to 48,000 in 2013 (Calle 2014). Between 2004 and 2018, the national government signed sixty-seven contracts with nineteen companies for the exploration of the oil reserves that the National Hydrocarbon Agency (ANH) estimates to exist in the Caguán-Putumayo sedimentary basin.[8] Thirty-seven of these contracts overlap with eighty-one indigenous reservations (*resguardos*), principally in the departments of Putumayo and Caquetá (Asociación Ambiente y Sociedad 2019). In 2016, in anticipation of the signing of the peace accords between the national government and the FARC-EP, the president of the state oil company, Ecopetrol, was quoted in the newspaper affirming that "peace is going to permit us to extract more oil from the zones that were prohibited by the war. . . . With peace we hope to have the possibility to enter into Caquetá with much more force, Putumayo, Catatumbo, areas that were difficult to access."[9] Colombia's Law 160, passed in 1997, prohibits the titling of land within a five-kilometer radius of oil or mining activity. Rural communities argue that this law contributes

FIGURE 3.3
USAID postcard
distributed in
Putumayo during
the 2009 Volvamos
a la Vida campaign.
Photograph by author.

to the ongoing concentration of land and the precarity of property titles in a country that has never undergone a full labor incorporation period, a history of agrarian reform, or a populist phase that even temporarily separated the state from traditional national elites despite Colombia's reputation as one of Latin America's most stable democracies (Carroll 2011).[10]

The prevailing analysis among communities living in coca-growing regions is that antidrug policy has created the conditions for an intensified mode of transnational capitalist expansion, linking the role of foreign direct investment in war to the securitization of development. A concrete example of this securitization was the creation of approximately eighteen Special Energy and Road Battalions whose sole mission is to militarily protect the mining, energy, and road infrastructure in the country (i.e., pipelines, drilling rigs, power generators, and the roads traversed by oil tankers and commercial tractor trailers).[11] Along with waging a war against the people (*guerra contra los pueblos*)

FIGURE 3.4 "Killing for religious motives is savage. The civilized way is to kill for economic motives." Cartoon by Eneko.

(Paley 2014), antinarcotics policy has become a pretext to wage a war against life (*guerra contra la vida*)—more explicitly, a war waged in defense of capitalist growth at the expense of all forms of life. In the process, a variety of social values become reduced to one exchange value, and a diversity of practices that are said to produce death in life (i.e., illicit coca cultivation), or that cannot be assimilated to or that openly resist growth-oriented imperatives, are actively restricted and eliminated. Rural communities in Putumayo have perceived militarized aerial fumigation as yet another violent attempt to weaken their will—a mode of destroying the material base and food crops for local sustenance and starving them out in an effort to force the abandonment of territory that facilitates industrial oil and mining concessions. Paramilitary repression, criminalizing public policies, and the militarization of daily life are central components in a war that declared itself to be against communism and narcoterrorism while obscuring its principal economic aims.[12]

As I wrote the first draft of this book, 6,300 members of the FARC were engaged in what they called their historic "final march" to the transitional zones where they demobilized and began the legalization process for reentry into civilian life.[13] This was poignantly described by the former commander of the Eastern Bloc, who wrote, "day-by-day we [guerrillas] convert into that which we once were: civilians, campesinos, workers, the nation's poor."[14] Unsurprisingly, the FARC's civilian and political transition has not been an easy or only celebrated reality. With the presidential election of Iván Duque in 2018, there is ever more concern about the government's ability to uphold and honor the peace accords. Not unlike the case in the rest of Latin America, the selective assassination of hundreds of popular leaders, especially trade unionists, environmentalists, human rights defenders, and protectors of land, water, and territory, has continued during the country's "post-accord" transition, along with the assassination of demobilized FARC militants and their family members.[15] Many doubts exist about the viability of the new National Program for the Integral Substitution of Illicit Crops, and coca growers throughout the country have reported the ongoing state application of repressive eradication strategies, including the manual application of glyphosate and the Duque administration's attempts to constitutionally reinstate the aerial fumigation of glyphosate. Simultaneously, the government signed twenty-eight Regional Pre-Accords for the Substitution of Crops of Illicit Use with regional social movements in the major coca-growing regions, such as Putumayo.[16] Coca, marijuana, and poppy growers and workers have organized into a national coordinating body, COCCAM, to demand that they be treated as political protagonists and legitimate interlocutors in a process to build profound structural changes in antidrug policy.[17] Many rural communities are impacted by the territorial power vacuums left behind by the demobilized fronts of the FARC, which are being filled by nefarious actors and criminal networks, dissidents, and reemerging paramilitary groups. Even government officials recognize the important role that FARC's presence played in protecting and policing the country's remaining primary forests and biodiverse corridors.[18] In 2015, one year before the peace accords were signed, 124,035 hectares were deforested, according to the Institute of Hydrology, Meteorology, and Environmental Studies (IDEAM). One year later, this statistic increased to 178,597 hectares—a 44 percent increase in deforestation—and 2017 recorded a devastating 219,973 hectares of forest eliminated.[19]

There has been growing public debate and recognition in Colombia that forests, soils, rivers, *páramos*, wetlands, mangroves, selva, seeds, fauna, and biodiversity of all kinds may also be "victims" and scenes of war requiring reparations in the country's post-accord transition.[20] More than attempting to repair degraded "landscape units" or "natural resources," the construction of peace for and from territories that have been epicenters of war requires a relational approach—that is, an attention to ruptured socio-ecological relations across multiple scales and temporalities, and to the loss of the capacities of communities to remain and flourish in their territories due to the destruction of the material conditions for work, food production, collective autonomy, and cultural reproduction. Indeed, this may be the decisive crux between the perpetuation of war that goes by another name—war that does not recognize itself as war—and the possibility for what many sectors of Colombian civil so-

FIGURE 3.5 Guerrilla troops from the Fronts 32, 48, and 49 of the FARC-EP Southern Bloc traveling to the transitional zone La Carmelita, Putumayo, to demobilize in February 2017.

ciety have demanded as "peace with social justice from and for the territories." War by other means is not only characterized by the reconfiguration of armed actors, dissident groups, and criminal networks, and the perpetuation of the assassination of social leaders and defenders and guardians of territories. This is a war waged by the courts when they render illegitimate the constitutionally recognized rights of municipalities to prohibit extractive activities in their territories and when they allow mayors to be sued for signing these municipal accords. It is a war waged by the subtraction of areas of forest reserves, modifying the use and vocation of soils to allow for oil and industrial mining concessions, and when the Land Restitution Unit (URT) and other state entities restore forested areas to victims, which changes the future environmental determinants of a place and renders possible the eventual arrival of extractive industries and other industrial activities. It is difficult to propose peace with social justice if conceptions of the social are informed by a modernist ideology that separates life into ontologically distinct categories: nature and culture, subjects and objects, bio and geo, living and dead. How might processes of justice seeking and making be transformed if violence and dispossession are treated as shared, albeit uneven, experiences of a multitude of beings and elements that compose and decompose into and thus make a given place or territory?

It is specific conceptions of *life*, along with human rights or land rights, that have come to the fore in rural struggles against extractivism, GMO seeds, biofuels and other forms of industrialized agriculture, large-scale infrastructural projects, free trade agreements, and all sorts of neoliberal reforms privatizing public goods and services by fomenting transnational capitalist growth across the hemisphere.[21] Of course, development has always signaled more than just material progress and economic growth indicators: it has marked a historically specific model of judgment and control over life itself. Although modern capitalist principles such as growth, progress, better living, and their correlate—more development—have been thoroughly repudiated in theoretical debates emerging in the Global South since the mid-twentieth century, these principles remain politically and economically dominant (Escobar 1994, 2014). Indeed, struggles over the definitions of and relations to "nature" and "resources" have become one of the most salient features of contemporary Latin American political dynamics (see Blaser 2009; de la Cadena 2010; Escobar 2008; Graeter 2017; Li 2015; Ulloa 2017). Over the last almost fifteen years, the region has witnessed a growing tension and pendulous shifts between a forceful "leftist" resurgence and a wave of renewed conservative governments. Notably, both self-declared progressive and more conservative administrations rely on what some scholars have referred to as a neo-extractive development paradigm to fuel their convergent economic models and divergent ideological projects (Gudynas 2014; Veltmeyer and Petras 2014). This raises several questions: How can radical life processes and structural transformations come into being and sustain their existence if socio-ecological justice and sociopolitical inclusion tend to be propelled by extractivism? What relations to life, death, place, and territory will be potentiated and which others will continue to be criminalized, rendered obsolete, or sacrificed in the name of economic growth and social(ist) good?

CULTIVATING OJOS PARA ELLA

[Eyes for Her]

Traveling toward the municipality of Santa Rosa at dawn reveals panoramic views of the eastern Andean foothills interrupted by military tanks and soldiers patrolling the sides of the highway. As we dip around a curve overlooking the expansive Río Cauca, we meet the same four tanks that have been stationed here for months. It is August 2010, and I am accompanying Heraldo Vallejo on a visit to a farm. He asks aloud why the soldiers keep parking on the same bend, leaving themselves vulnerable to mortar fire by the guerrillas. We peer out the truck's windows, and what look like thumbprints pressed into the side of the mountain expose patches of rust-colored soil interspersed between low-hanging clouds and dense selva canopy. Heraldo tells me these are scars left behind by heavy rains that rip up tree roots and send chunks of earth crashing down the ravine. Naked soils are then left to bake beneath an intense equatorial sun. This entire area is designated an "unstable geologic zone," as the highway sign warns. Ironically, a plastic banner hangs next to it, stating: "Travel with confidence, your army is on the road." One of the other passengers in the truck makes a comment on the "insecurity of security," and as we leave the department of Putumayo and cross into a neighboring subregion referred to as the Media Bota Caucana, we find ourselves squarely in the midst of the transitioning piedemonte-amazónico (Andean-Amazonian foothills). The resulting shift feels like a hundred microecologies with their tiny alterations in temperature and diverse plants and trees that hover above wetter, heavier soils.

Today we are visiting a campesino named Edelmo to discuss a proposal that he and his family designed for a new farm that they recently acquired. Having defaulted on a loan on their farm in the neighboring Andean department of Nariño, they migrated to Santa Rosa, Cauca, in 1999 to follow the coca and illegal timber booms. After extracting cedro, *guamo*, and *sangretoro* trees—grueling work that leaves both man and mule as good as dead, local people often say—Edelmo saved enough money to grow a twelve-hectare seedbed of coca that he sold for $2,500, which was sufficient to invest in a farm with three of his brothers. However, intensified aerial fumigation and the widespread arrest of people found with gasoline, cement, or bulk food assumed to be destined for the leftist insurgency led Edelmo and a group of his neighbors to organize the voluntary eradication of their coca plants. Now he was back to coffee, plantain, and subsistence farming, which had been his family's previous vocation in Nariño.

"We kept looking for the soil, those quality soils we were accustomed to seeing in Nariño. After eight years experimenting on the farm, we are still learning how to practice permanent agriculture here," he tells us as we hike past coffee bushes intermixed with fruit trees, a few vegetables, a row of henhouses, and pens of guinea pigs. Edelmo is the leader of a group of thirty rural families interested in growing what they call ecological Amazonian coffee. After the municipal secretary of agriculture failed to support the initiative once he learned that it did not involve raising cattle, Edelmo spent a month hiking up and down a nearby hill with his cell phone to catch the spotty reception that allowed him to call the National Federation of Coffee Growers (FEDECAFE) in Bogotá. However, FEDECAFE kept telling him that he was not located in a coffee-growing zone. "They say it is the wrong thermic floor with the wrong class of soils, and they refuse to come out to see how far we have come along or to even taste the quality of the coffee," he told us, visibly frustrated. Furthermore, the regional office of what at the time was called the Colombian Institute for Rural Development (INCODER) kept denying him a land title for the new farm that his family is planning across the highway. Edelmo solicited a title for sixty-nine acres of largely primary forest, forty of which he intended to remain as forest and twenty-nine that he would exploit agriculturally. When asked why two-thirds of the area would not be "improved" (i.e., cleared), he explained to the INCODER officials that he plans to incorporate the protection of natural forest into his agricultural system. However, the agency rejected his application and instead issued him a title for what they considered to be an economically viable "working farm." Edelmo questioned seemingly contradictory state development and environmental policies. On the one hand, government officials argued in favor of forest conservation and cracking down on an "unorganized" and largely spontaneously expanding agricultural frontier. On the other hand, national development plans actively partitioned the territory into concessions for oil exploitation and industrial mining, crop-duster planes zoomed overhead spraying swaths of selva with glyphosate, and the military bombed suspected insurgent camps and in the process obliterated everything else in sight.[22] As we head down a slippery bank, Edelmo sheepishly acknowledges the existence of a row of coca plants that he tells us stubbornly rear their heads whenever he looks the other way.

Despite the complications with INCODER, the family says they plan to forge ahead with their life project. In the kitchen, Edelmo unfolds a hand-drawn map of the new farm, Melina, named after his youngest daughter, and tells Heraldo that he invited him today because of his reputation as el hombre amazónico. Before Edelmo goes on to share the details of the agroforestry project, Heraldo suggests conducting an inventory of the current farm. He asks them: What does the family buy? What do they produce? What is their relationship with the selva? After making a list, the balance unequivocally tips to one side. They buy just as much as they produce and eat only five "wild plants": *cilantro cimarrón*, passion fruit, custard apple, the *mil pesos* palm, and something that looks like a cherry but they are uncertain of its name. Openly admitting their unfamiliarity with much of the plant and tree life growing around them, they tell us they never fish in the river and do not have a garden, conventional or otherwise. Heraldo goes on to say, "We have this idea that we have to eat what we plant and if we don't have a place to plant, then we buy food or go hungry. Agronomists tell us that to farm we have to 'correct' the soil and grow high-yield market varieties. They say that any other way is backward, and now we are the ones who have become poor, dependent consumers. What if cultivating here means cultivating ojos para ella [eyes for her]—la selva?"

As the months passed, I would learn that cultivating ojos para ella relates to recovering and innovating a whole series of practices that extend far beyond human-oriented production and consumption. Furthermore, "not having a place to plant" did not necessarily refer to a material or physical lack, but to ruptured relationalities, which are also relations that had yet to be cultivated and transformed. Edelmo and his family expressed an uneasy estrangement from the dynamic and myriad associations that compose and decompose the place in which we were physically standing, and thus from the world of which their lives form a part. During the ride back to Putumayo, Heraldo told me that neither Edelmo nor INCODER are enredados (entangled) with la selva. Edelmo is hesitant because when he looks out beyond the limits of his crops, he sees a strange world of rastrojo (weedy regrowth), while INCODER is unable to see anything but a physical space that should be "protected," "policed," or "worked" (i.e., hectares of land). Cultivating eyes for her, la selva, is not the same as only having eyes for her. It is not a possessive gaze of capture, although seduction may be involved in attuning one's senses to different textures, tastes, and uses that are part of learning to see and inhabit the world differently. Indeed, perhaps selva is not even a woman at all. Instead, I think of her as the spirit of that which is to come, of that which was, of selva thought and memory—a memory that is not trapped in modern dualisms, such as masculine and feminine, bios and geos, or sentient and inert.

During our next visit with Edelmo, we head across the highway from the family's current farm. Hiking up a slick path, the familiar heads of palm trees and ceibas tower above less imposing varieties. Seeds litter the underbrush, along with fallen fruit, rotting rinds, layer upon layer of branches, and vines that twirl up and down and all around us. "What do you see?" Heraldo asks us, and before we can answer he says, "I see a salad." He smacks his lips and I imagine him digging into the array of meals that begin to surface before our eyes. Edelmo and I quickly realize that we inhabit a place that is at once the same, but very different. Weeds become salad, tree bark yields pungent spices, and a dozen tubers, nuts, fruits, and vegetables emerge out of corners of the woods. A trail of leafcutter ants drags a meal underground. We steer clear of fist-sized venomous spiders as we pass tiny waterfalls only to have our

FIGURE 3.6 Harvesting tubers in the Media Bota Caucana, August 2010.
Photograph by author.

necks nipped by fire ants on the other side. Heraldo stops to gather *buchón*, an
aquatic plant that absorbs heavy metals, and suggests that Edelmo sow several
rows near the family's drinking water. He then begins to point out a pattern
of leguminous trees such as guamo, which fix nitrogen, palm trees and ferns
that concentrate phosphorus, and waxy-leaved trees like plantains that pump
potassium to an underground invisible world: "You have to chemically fertilize
with synthetic 10-30-10 to create accessible N-P-K," Heraldo says, performing
a brief caricature of the agronomists who accompanied different phases of the
illicit crop substitution and alternative development projects funded by Plan

Colombia.[23] "We say no. NPK is all around us, created by communities of plants and microorganisms. They feed each other while they nurture the ground. If you don't trust me, at least reach an agreement with la selva. She won't allow for impositions," he tells us.

During these farm visits and the alternative agricultural popular education workshops that Heraldo was invited to attend and lead, and when I spent time with him on his own farm, I observed processes of unlearning and relearning, or what I came to think of as diverse trajectories of becoming a *selva apprentice*. I understand apprenticeship not as an environmentally deterministic submission to, but rather an experimental process of learning to *follow* la selva, that is, to cultivate and be cultivated by selva.[24] Stengers and Pignarre (2011) speak of "trajectories of apprenticeship" to refer to heterogeneous and divergent processes that are always situated by the place in which they have been able to gain a hold. They argue that trajectories create not an image of a massifying movement, but instead "new means of grasping a situation, leading to the production of new ways of acting, of connecting, of being efficacious where the classical protagonists had accepted the problem as formulated by a supposedly scientific and hence 'neutral' [and generalizable] expertise" (55). The kind of unity that emerges between different trajectories of apprenticeship does not produce a feeling of being "in common," but a capacity to, in their words, "vibrate together" (54) and to draw from this vibrating energy new capacities to imagine and situate oneself within a particular milieu.

Heraldo and I often joked about what I liked to call his selva glasses. However, more than a lens that rural families strap on to fine-tune their gaze of tropical forest ecology, cultivating ojos para ella (eyes for her, la selva) is an ongoing un- and relearning process that occurs in all of one's senses—from fingers to feet, heart to intestines, and eyes to tongue. I conceptualize this less as an emergent environmental subjectivity and more of a transitioning and ongoing relationality. Cultivating eyes for her helps the Amazonian human learn to walk differently, experience new tastes, identify forgotten seeds, cultivate "weeds," recycle "waste," hear plants, sense vibrations, and exchange differently. The Amazonian human is one in which the human—what it means to be human—necessarily becomes composed by and decomposes back into selva when one follows "her" without the guarantees of preestablished human mastery or human-only sensory field. In other words, "cultivating eyes for her" produces a different kind of human, a human that becomes one with selva's agroecological and territorial conditions rather than establishing one's humanity, and hence ontological singularity, by exerting dominance over and colonizing her. I met families and rural networks of campesinos that had begun to recover noncommercial seeds and to experiment with new recipes, re-place Andean potatoes with autochthonous Amazonian tubers, and cultivate communities of shrubs and trees that generate their own N-P-K. Some of these families collect human and animal urine and feces to nourish microorgan-isms and plants that, in turn, help prepare food for humans and other animals. Others design crop layouts following the direction and intensity of the sun, producing multiple layers of foliage that, in turn, produce soils that are always disappearing in the process of becoming other organisms' nutrients. They recommend replacing fenced-in, ground-level gardens of temperate Andean vegetables (conventional lettuce, tomatoes, and onions) that require constant tending, chemical inputs, commercial seeds, and replanting after each harvest. They instead grow Amazonian orchards and gardens where creeping plants and tubers roam, as they say, "free," and produce food for humans and other creatures every fifteen days. Simply put by Heraldo, although not at all simple in practice, a person who wants to avoid becoming dependent upon externally dictated market dynamics and conventional agronomic models must in turn avoid trapping plants, seeds, soils, and trees. However, no new, singularizing agricultural model exists—only seedlings, experiments, and lessons—to be shared and refashioned or not from one farm to the next. What joins these families and campesino networks is their desire to create alternative ethical-material landscapes with their corresponding economic, political, and ecologi-cal transformative possibilities. Questioning what it means to define a "soil" as "good" and "productive" provokes questions about the values and practices that underpin a market-oriented and ultimately human-centered notion of productivity itself.

I place Heraldo's proposal of cultivating ojos para ella in conversation with what Stengers (2005c) calls an "ecology of practices." In this situation, par-tially connected practices for working, growing, eating, shitting, and decom-posing that need each other because none of them offers a total response for how to follow and respond to la selva. Each time a practice is passed on from one person to the next, one flux of microorganisms-air-water-seeds-sunlight-hens-rootlets to the next, this practice is reshaped and rejustified. Stengers re-minds us that practices cannot be separated from their milieus, and when ap-proached with attentiveness to how they diverge in response to each situation and the questions and concerns relevant to these situations, they may enable the creation of what she calls a "different practical landscape" (2005c, 187). Cul-tivating eyes for her entails learning how to engage in processes that Heraldo conceptualizes as *lecturaleza*—*lectura* (reading) *naturaleza* (nature)—or what I translate in a similar play on words as "readinature." Other small farmer friends, Nelso and Elva, call this a process of *ojimetría* (eyemetering) that relies on skillfully cultivating one's visual field and the art of touch and tact.

I do not understand lecturaleza as something akin to biomimetic reproductions of "nature." The field of biomimicry is deeply tied to US militarism and the technological developments of the military-industrial complex. Furthermore, lecturaleza does not depend on an ideal of an object "out in nature" that is abducted back to the creative intentions of a human mind, which is somehow external to what is being reproduced. Neither does lecturaleza necessarily depend on variables that can be fixed and held constant across differing conditions. For example, Heraldo would advise me that the best way to avoid consuming poisonous plants is by following cows as they graze, or that if I ever get lost in the selva I should follow the movements of monkeys from tree to tree. He taught me that lecturaleza implies an apprentice-like following, and hence joining in and replying to the flow of forces, beings, and elements composing and decomposing a particular place. "Following," as Deleuze and Guattari point out, "is not at all the same thing as reproducing" (1987, 372). Whereas reproducing involves a procedure of reiteration, following involves itinerancy and ambulation—not to describe the relay of relations between one thing and another, but rather the act of following the lines of growth and movement along which things continually come into being and unfold. Deleuze and Guattari insist that whenever we encounter matter, it is matter in motion, in variation and transition, and thus "matter-flow" (409) can only be followed. Similarly, lecturaleza is not a question of people copying what cows or monkeys do, but, as I understand it, the itinerant and improvisational work of following cows, monkeys, insects, rootlets, and solar and lunar cycles within changing material conditions and emergent situations: nutrient cycles, bacterial reproduction, the decomposition of fallen leaves, rainfall patterns, and shifting humidity and climatic factors.

Itinerant apprenticeship of transitioning relations—such as the growth of plant from seed; an animal as it eats and shits manure that decomposes into the metabolic cycles of microorganisms that feed plant rootlets, which are also fixing nitrogen from the air; the differing durations of direct sunlight in a given place; and the rotating phases of the moon—produces what Heraldo refers to as *conocimiento vivo*. Living knowledge emerges from the trajectories of apprenticeship that constitute one's life and labor, and the many lives one labors along with, follows, eats of, and defecates back into. He contrasts this attentive and experimental following and responding to lecturaleza with the scientific objective of *cosechar conocimiento* (harvesting knowledge). Lecturaleza is not simply a process of trial and error, but a process that emerges from the necessity to solve concrete problems and ask questions that are relevant to daily life on the farm and one's territory.

Many rural families that migrated and/or were displaced to the piedemonte amazónico, such as Edelmo's family, express problems similar to those of IGAC agrologists. They seek to locate and work a "productive" and "quality" soil, and when this soil cannot be located, they turn to a specific range of solutions that include fire, synthetic fertilizers, cattle, and abandonment. It is precisely the ephemeral and fragile constitution of these local soils that is claimed to thwart human projects. However, for Heraldo and the other families participating in the diverse alternative agricultural networks I came to accompany, the problem to be solved is not the same, and thus taxonomic classifications of soils intended to determine their "productive capacities" do not offer solutions to the "enigma" of the Amazonian plains. These families do not contend with soils that are "poor" or even soils that are "different." Soils are a relation into which they also disappear and become something else that they also compose.[25] Even when these rural practitioners interact with state officials, they do not often articulate "soil" as a stable object, but rather as a bundle of relations that do not leave anything behind—not pests, urine, weeds, or even their own feces.

LECTURALEZA

On one occasion, I joined Heraldo at a meeting with an association of aspiring coffee growers in the vereda Verdeyaco in the Media Bota Caucana. One of the campesinos stood up to speak on behalf of the group, and he told us that the *entorno* (surroundings) are quite distinct from the places that many of them called home and were forced to leave behind in the Andes. He and his neighbors "feared clashing with the territory," he said. They wanted to avoid converting it into a desert of monoculture coffee crops. "We do not merely want to be economically sustainable, but also ecologically so," he explained. In the absence of appropriate Andean-Amazonian technical assistance on the part of the state, no one in the community was sure how to proceed in designing their polyculture coffee farms.

"Who should we ask?" Heraldo posed to the group. Someone suggested that it would be best to consult the longest-standing inhabitants of the region since they must have figured out how to adapt to "such harsh selva conditions." While this was not considered to be a bad idea, Heraldo had something else in mind. "How about the plant itself? Why do we usually think that only outside experts can give us technical support?" he asked. "Why do we think that only human minds know?" He then continued: "The farms talk, the plants know, and the human family replies." I return to the idea of "talking farms" in chapter 5. To illustrate his point, Heraldo shared an anecdote with the group. The story went like this: A farmer is hard at work one day when a pair of state agronomists arrive at his farm raving about a new and improved hay seed they are distributing throughout the region. They tell him that this seed is the best technology yet, and that it is sure to double yields and of course profits. He should start sowing it right away. Before departing down the road to the next farm, they generously hand him a complimentary bag of seeds. This campesino scratches his head and takes a minute to think the situation over. He says to himself, "Who should I consult about this seed? If I go talk to another agronomist he is going to tell me that this new variety is a surefire invention. If I go to the municipal secretary of agriculture they will try to convince me to plant God knows how many hectares. If I ask my neighbor, he will probably just tell me that he doesn't know either. Who should I ask?" The man was puzzled for another minute and looked around his farm, but then seemed relieved. "I know. I will ask my mule." He then planted two seeds—the new variety and his traditional one—and when it was time to harvest the hay, he placed both varieties on the ground in front of the mule. "Now you choose," he said. The mule walked toward the hay produced by the new seed, bent his head, gave it a sniff, and turned away in disinterest. He then went over to the farmer's traditional variety and sat down to eat. "Well, I guess I have my answer," the farmer said, and he swore that the mule looked up and gave him a wily glance.

STATE SOIL SCIENCE: CLASSIFIED WITHIN CLASSIFICATION

As we trek across a muddy field in the municipality of Fusagasugá in Cundinamarca, two hours outside the capital city of Bogotá, an elderly woman leans over her patio railing to offer us a cup of coffee while Oscar, the agrologist I am accompanying, explains that her property is part of a pilot zone, which IGAC has deemed representative of the dominant relief and climate in the municipality. He asks if she would mind if we stop *para levantar* (to lift) some soil samples from her farm. Immediately, the woman appears nervous, and she assumes that our visit is related to land valuation for state taxes. However, once Oscar assures her that the samples will be used to determine soil vocation for the municipal development plan and not directly affect the family's property taxes, she waves us on our way. Oscar and two men from Fusagasugá who have been hired for the day begin to dig a sixty-centimeter-deep cavity, revealing what scientists call a "soil profile" that produces both an object of

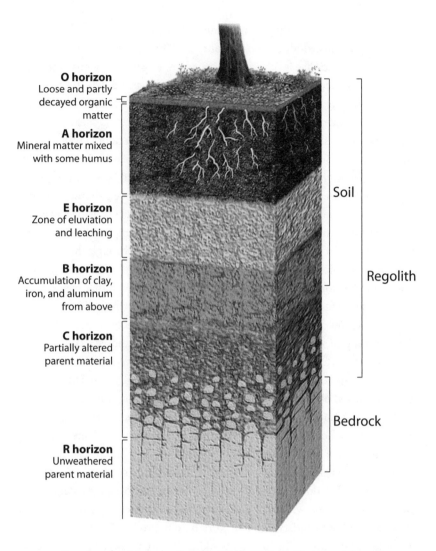

O horizon
Loose and partly decayed organic matter

A horizon
Mineral matter mixed with some humus

E horizon
Zone of eluviation and leaching

B horizon
Accumulation of clay, iron, and aluminum from above

C horizon
Partially altered parent material

R horizon
Unweathered parent material

Soil

Regolith

Bedrock

FIGURE 3.7 Drawing of a standard soil profile (Wolfe 2001, 9). Photograph by author.

study and a sense of pleasure as we admire the diverse colors and textures of the world previously hidden beneath our feet. This vertical cut exposes what in standard soil science is referred to as the o, a, b, and c horizons with their corresponding depths, hues, and sculptures. It is in such details that a soil is said to express its "personality" and through which soil scientists begin to interpret the "soil's speech."

From cow pastures, stony hillsides, and mossy forest floors I watch as soil samples become such—as what to me looks like mud—leaving behind a thick existence (locality, historicity, ecological relationality) to be sealed in plastic bags for the two-hour journey to the National Soil Science Laboratory in Bogotá. While in the field, Oscar begins a classification process using a Munsell color chart, which translates levels of moisture. For example, he tells me that slate hues suggest water may not be channeling freely through the soil horizons. With a simple lens he analyzes pore size and counts rootlets and uses a hand kit to test for the presence of sponge-like volcanic ash that may signal a reserve of water is underfoot. Texture is categorized in terms of the percentage of sand, silt, and clay content. He then adds a few drops of water to different pieces, rolling them into tiny balls in the palm of his hand. Sticky, greasy, or coarse is then jotted down under the category of consistency on his clipboard. Before moving on to the next dig, Oscar records the vegetation, land use patterns, relief, and other visible characteristics of the surrounding terrain.

In Latour's (1999) well-known account of soil scientists at work in the Brazilian Amazon, he conceives of the Munsell color chart as an "intermediary" that forms part of a successive chain of transformations and inscriptions that allow a bounded entity to be abstracted. None of these intermediaries resemble anything; they do more than resemble. As Latour notes, they stand in for the original situation without ever substituting completely for what they have gathered.[26] What scientists seek and work hard to produce is a portrait of soil as a unique individual with an inner working that is considered to be reflective of its "nature" and "personality": it is always a kind of soil; no generic category exists for this entity. The soil as a vital nexus reduced to increasingly simpler parts that can then be expressed in the language of statistical equations also defies this summation of parts and resists complete dismemberment into a set of isolated essences. When the earth beneath our feet is uncovered in the act of making a soil profile, I witness the first movement of observation, substitution, and "boundary-drawing practices" (140). Clumps of earth begin to be bound up in words, numbers, symbols, and graphs that enable the passage of dirt and muddy clumps to text. At the same time, the career of soil scientist is fashioned, and the soil has changed states in order to change locations.

FIGURE 3.8 Hellige-Troug Soil Reaction (pH) test conducted by Oscar in the field during the Fusagasugá soil survey, August 2009. Photograph by author.

When we arrive back at IGAC, bags of dirt are sent to the laboratory and divided up among different scientific disciplines. I watch microbiologists separate out dead plant roots from living organisms before they place the bags in a refrigerator to retain its humidity. They struggle to ensure that what was once teeming with life continues to harbor a measurable quantity. Soil physicists weigh the samples and leave them out to dry on the rooftop, sifting and weighing them again once all the water has evaporated. However, this may take weeks, depending on Bogotá's infamously fickle weather, and work schedules are dependent on the relative humidity, hours of sunlight, and levels of precipitation on any given day. Chemists receive what by now is a powder-like substance and proceed to dilute it even further into test tube–size solutions to conduct experiments on levels of fertility. As Oscar explains, each time a humus colloid is disaggregated into its most basic chemical components, scientists arrive at an acid of a slightly different concentration. Humus demands that the same experiment be run six times to coordinate the differing results. What Oscar later interprets as humidity, pH level, fertility, and clay and organic content is a partial re-representation of relations that were necessarily severed in the journey from field to color chart, laboratory, refrigerator, and

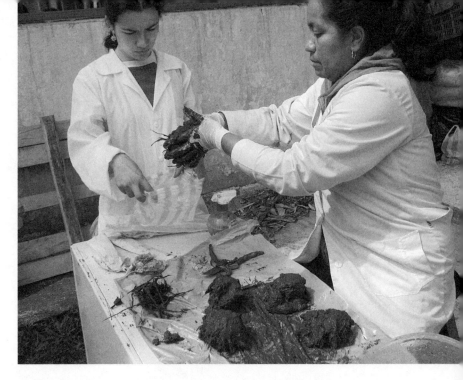

FIGURE 3.9 Sorting soil samples as they arrive at the National Soil Science Laboratory. Bogotá, November 2009. Photograph by author.

test tube solution. It is what allows him to define the sample we lifted from Fusagasugá with precision as an inceptisol: young, acidic, with low fertility, 50 percent chance of susceptibility to erosion, and of limited agricultural use.

Furthermore, this classification of soils entails a comparison of a specific kind, in which a definition of what is "fit/unfit," "stable/unstable," "functional/ dysfunctional" organizes the things compared in a hierarchy. From a scientific perspective, soils are defined within a taxonomic system based on their "natural" properties; however, as I outlined earlier, taxonomic orders in themselves are not of interest to the state. What is of interest is the additional sorting of these groups into technical classifications for potential soil use. In Colombia, both classification systems were officially adopted in the 1970s from the US Department of Agriculture (USDA): soil taxonomy (twelve orders) and land classification capabilities (eight classes).[27] The latter hierarchically organizes soils according to limitations that have an impact on their use for the production of conventional crops and pasture plants without deteriorating over time. Class I, II, and III soils are categorized as slightly restricted for agriculture, while VI, VII, and VIII are deemed severely limited, unsustainable for cultivation, and restricted to grazing, forestland, wildlife, or aesthetic purposes. Even when mainstream scientific views of soil quality and health have begun to regard soils through a more ecosystem-oriented conceptualization (i.e., USDA-NRCS 2010), agrologists are expected to scale up from a portrait of a particular soil to the level of units of land based on a classification system embedded in capitalist imperatives that valorize agriculturally productive soils over all others. Thus, soils first become an object through their separation from land as the terrestrial ecosystem. However, their definition at once feeds back into land use management in its dimension as property and workable acreage with a potentially productive capacity. Following Bowker and Star's (2002) relational understanding of infrastructure, soils are also that which emerges in between—between what scientists define as "natural properties" and "social relations" mediated by technologies and emergent with differential value within tax codes, loan approvals, development plans, and, as Edelmo shows us, the (im)possibility of engaging in specific state-sanctioned agricultural life projects, and more profoundly, certain dreams of viable territorial transformations and alternative rural economies. It is in this double process of producing soils whose value depends on an ability to be translated into agriculturally productive units of land that IGAC agrologists employed to work in the Amazon become perplexed. Indeed, they find themselves confronting new problems that result from the very solutions posed by their classification systems.

As I learned with Heraldo and others, the impermanence of soils in the Amazon—their metamorphosing existence—makes them unable to be extricated from an entangled web where every element is implicated in the existence of the other. Thus, they come into existence paralleling what Karen Barad has called *intra-action* or the "mutual constitution of entangled agencies" (2007, 33).[28] How does one taxonomically classify an intra-action? State soil scientists are faced with the dilemma of how to render the land productive when no land can exist without a soil that is intimately at one with and feeding back into selva. The engendering of "soil" as a stable entity versus living with "soils" in perpetual relation is an inherent contradiction in the classificatory notion of what is or is not productive land. I watched this complex relationality remain a site of struggle for agrologists as the material and ethical force of scientific classification came to the fore.

In the 1980s, Cortés and a small group of Colombian agrologists attempted to mediate this tension by describing Amazonian soils as "different" rather than simply "poor" or "deficient." However, this "difference" is embedded in an ontological framework of comparison in which definitions of soils are based on natural conditions in temperate zones, such as depth, chemical fertility, neutral pH levels, longevity, and youth, as well as on moral codes delimiting their legal and proper use. The Western scientific terminology of a neutral pH is a value-laden term with ethical and semiotic charge given that slightly alkaline conditions are considered to be the most optimal. "Difference" inherently implies a deviation from a productive standard, which traps scientists in a hierarchy where "different soils" almost inevitably become problematic ones. Agrologists contend with an "elusive" terrain that troubles the limited reach of their inscription devices and the productivist economic rationale and temperate climatic and agroecological assumptions underpinning these devices. Indeed, when I conversed with an agronomist and entomologist who works for the FAO and IGAC, and who represented Colombia in the South American Alliance for Soils meetings, we discussed the other options that Colombian soil scientists could have pursued as an alternative to adopting the USDA soil classes. She told me that Brazilian scientists, for example, created their own taxonomic system given that a large portion of the country (i.e., the Amazon) was deemed unproductive according to the USDA system. In her opinion, IGAC would have been better off implementing the FAO's integral agroecological zoning approach. This coincided with Pedro Botero's perspective in the 1970s when the institutionalization of the USDA taxonomy occurred. However, she also thought that it would now be very difficult to switch classification systems because the entire country had only recently been surveyed at the most general scale of 1:1,000,000 utilizing the USDA system.

Bowker and Star (2000) invite us to consider classification as a work practice in which individuals make decisions that lead them to use categories in which they may not wholeheartedly believe and may find ethically compromising. Interestingly, it is Cortés, credited with institutionalizing USDA soil classification in Colombia during his tenure at IGAC, who would also be one of the first to publish articles questioning its universal applicability—particularly its unbridled development-oriented imperatives in the country's tropical forests (Cortés and Ibarra 1981).[29] In contrast to an "Andean mentality," he published articles urging his colleagues to assume a "conciliatory stance" toward the Amazon by imagining agricultural systems that would both care for soils and guarantee their "rational and productive use." More than a mentality, this Andean nation-making lexicon refers to a historically informed concept of production that is premised on separation from a nature that is seen as both object and patient, and that acquires value once its "inherent limitations" are overcome through chemical inputs and other commercial amendments and human inventiveness. In reference to the Putumayo soil survey, Cortés told me, "We should make recommendations that totally change soil use and management instead of forcing the region to fit into USDA taxonomies. It would be excellent if we could recommend agroecology or agroforestry." However, IGAC agrologists have found their attempts to produce what I call a "minor science" constrained by their classification systems, institutional positions, funding sources, and status as technical advisers and not policy makers.

I found a similar situation in the applied agricultural microbiology laboratory of IBUN. When I was asked early on in my fieldwork to give a presentation about my ethnographic research within the space of the weekly lab meetings, the director of the laboratory, Daniel Uribe, added a corrective to my characterization of the lab. He said: "Your presentation gives me the impression that you think that we are the other reality of the industrial associations. Unfortunately, we are not providing an alternative technology for the agricultural management techniques of the industrial guilds. Our work with the National Rice Federation [Fedearroz], for example, is temporary and conjunctural. In two years, we will probably no longer be working with rice." Uribe disagreed with the level of power and influence that I had attributed to the IBUN soil microbiologists given their collaborations with industrial agricultural associations and their funding support from the Ministry of Agriculture and the Administrative Department of Science, Technology, and Innovation (COLCIENCIAS). He and the other microbiologists in the lab emphasized that their collaboration with Fedearroz occurred only after the rice industry suffered economic setbacks and realized that they would be forced to directly compete with the cost of US rice imports after the protective conditions built into the free trade agreement signed between the two countries expired. "Once they realized how it would affect their wallets, they came looking for us," Javier explained. "They said, didn't you talk to us about some microorganisms before?" The industrial guild's attention to applied soil microbiology was hitched to international trade agreements and capitalist flows and fluxes, and not, as the IBUN scientists argued, the rice sector's sudden transformation of their myopic focus on seed improvement and genetic manipulation to consider the potential beneficent role of soil biota.

As I mentioned at the opening of the chapter, IGAC agrologists admit that while state soil surveys imply fieldwork, they do not entail consulting with rural communities that have their own situated and conceptually informed material practices. For example, rather than categorize soils into two generalized "agriculturally unproductive" taxonomic categories, in 2002, Heraldo and two other colleagues outlined nine types of agro-land systems with their corresponding agricultural potentials and limitations: scarp, highlands, hills, meson plains, low hills, ridges, wetlands, seasonally flooded Amazonian *varzea*, and *vegas* produced by the sediment from Andean-born rivers (*escarpes, terrazas altas, colinas, mesones, lomeríos, vegas, varzea, cochas y humedales*) (Vallejo, Campaña, and Muchavisoy 2002). Agricultural vocations are not conceptualized as separate from wetlands, lakes, watersheds, forests, and surrounding rock outcrops. Neither are there "bad" or "poor" soils. In a 1993 article published in a local journal in Putumayo that no longer exists, Heraldo wrote, "the lands here cannot always be seen as 'soils' with arable layers. Neither can they be understood as substrates in need of the chemical correction of their acidity or as a deposit for nutrients to balance fertility . . . soils are living organisms, processes of decomposition, litter, and sunlight" (1993b, 18). For this reason, the more recently implemented government *Servientrega suelos* service—which I translate as "door-to-door soil analysis"—whereby farmers are encouraged to send soil samples to an urban laboratory through a mailing service and then ten days later receive a soil study and technical recommendations for the chemical fertilization for a particular commercial crop, are of no interest or use to most rural communities in Putumayo.[30] This kind of laboratory analysis consistently reports that local soils are excessively acidic and of poor quality.

To avoid rushing in and contributing to an inefficient history of land use and worsening rural poverty, IGAC agrologists preferred to postpone the publication of the Putumayo soil survey. I read this "standstill" as an active hesitation: one way that soil scientists manifested a moral dilemma or sense of nonconformity with the treatment that Amazonian soils receive within dominant agricultural paradigms, even when they author the studies that inform these public policies. Much as in the soil sampling experiments that the engineers engage in in Harvey and Knox's (2015) ethnography of highway building in South America, IGAC agrologists were acutely aware of the provisionality and limitations of their classificatory and land use recommendation work.

Heraldo first introduced me to Cortés's articles because he finds them to be a kind of scientific ally for rural communities in the Amazon who are attempting to resist participating in extractive-based agricultural systems. Cortés's call for a "conciliatory stance" with the region opens up the possibility for conversations that question the transformation of selva into unsustainable farms—if not between state officials and campesinos, then among campesinos and alternative agricultural practitioners. However, Cortés's proposal also falls short because it further emphasizes human actors who manage the land, and thus presents alternative agriculture as a mere technical substitution rather than an ethico-political stance or life proposal. A sustainable approach to soil management is not a radical enough proposal for rural families that do not seek to isolate, correct, and employ an entity for the sole production of human food and profits. By representing Amazonian soils within a USDA taxonomic order (mainly weathered oxisols and ultisols), IGAC agrologists have tended to produce a disappointing soil at the bottom of a hierarchy that almost negates its existence, because to exist with dignity, soils have to produce economic "outcomes" and not intra-active entanglements. The recycling relationalities of selva soil in the Amazon reveal the limits of development imperatives in which production is premised on a deep-seated divide between "nature" and "culture," the contemporary purpose of which is for the former to produce for the life of the latter—or more precisely, to be unilaterally eaten and consumed by the latter. As I go on to discuss in ethnographic detail, Heraldo and other campesinos engaging in what I refer to as dispersed selva agro-life processes pursue sustainability with the understanding that growing food for humans necessarily implies nurturing an array of organisms, beings, and elements that capaciously oblige—at times though their sheer recalcitrance, robust fragility, and indifference—rural familes to regain a relative autonomy from chemical companies, aid packages, and other imperatives of market capitalism.

WARNING: HIGHLY TOXIC EXPERIMENTS

Further Reports from Putumayo

A hundred years from now, you might
wonder how they turned the butterflies against us,
how the graceful flight of such creatures
came to circle overhead
like a flock of angry birds.
Wings grinding together,
the screech of metal contraptions,
these moving metal contraptions sucking
the life out of everything.
The leaves of banana trees,
hen feathers,
scraps of human hair,
even the mushroom caps that crept across our rooftops
(this so-called second experiment).

I cannot tell you
how we scuttled out of sight,
beyond the scope of satellite maps
mesmerized by dark swarming clouds
and the trails of half-eaten foliage.

We crept beneath the not so distant sound of helicopter blades,
and the buzzing motors of low-flying planes.
This when it became clear:
The only solution is to arm yourself,
leave, die, or figure something out.
Shouted the general at the market,
repeated the experts through a bullhorn,
whispered the biologist to her butterflies
just before they took flight.

I wrote this poem after I heard rural communities in Putumayo speculate that US-Colombia antinarcotics policy had introduced the release not just of herbicides, but also of biological weapons in covert experiments to attack illicit crops. They described swarms of black butterflies descending upon fields, and a pathogenic fungus, *Fusarium oxysporum*, contagiously infecting forest floors. However, the butterflies' larval offspring seemed to munch on just about anything besides coca leaves, suspiciously mirroring the way chemical spray drift from aerial fumigation most often killed staple foods, pasture grass, forest canopy, and even US Agency for International Development (USAID) crop substitution projects rather than targeted illicit coca, poppy, and marijuana plants. No official record of the implementation of biological weapons exists—only peoples' haunting stories of the sounds of the flapping wings of black butterflies and the silent encroachment of fungal spores.

4

DECOMPOSITION AS LIFE POLITICS

On Reclaiming and Relaying

We are invited to gather around a spiral of seeds that have been carefully arranged on the dirt floor of the *maloka* (ancestral gathering house). The earthy hues of auburn-, red-, taupe-, and lilac-colored seeds are framed by dandelions, and kernels of corn are woven in and around the outermost ring. An open coffer with burning incense has been placed within the next circle, and a flask of water on a bed of dark granular soil rests in the center of the spiral, which all the seeds loop toward and back out from in their symmetrical circling around one another. Next to the flask is a yellow candle that Antonio bends down to light to pay homage to the sun. Once the candle is burning, we form a circle around the spiral. Antonio begins *la mística* (which I roughly translate as ritual) to inaugurate the Fifth Encounter of the Network of the Guardians of the Seeds of Life of Southern Colombia and Ecuador: "Seeds are the axis, the first inhabitants of the territory. They are a symbol of resistance, of not disappearing like so many campesino and indigenous communities at the hands of politics. Red is the power of rebellion, creativity, and design, thinking how and from where to resist in the thick of the war. Black is mourning, memory, and strength. From the time we are born until our deaths, even in death, they intend to separate us from the land."

Someone else in the circle intervenes to request that we hold a moment of silence to honor comrades, friends, and family members who have been killed by different armed actors or that continue to be disappeared after decades of war. We briefly sink into our private memories of loss and then collective reflections about the various kinds of death that haunt, uproot, and sever rural life. Paramilitary violence has been most gruesome in the public dismembering of bodies and the disposal of corpses into local rivers, a kind of ultimate deterritorialization of bodies from soils and souls. This is a macabre death that violently extinguishes life, attempts to silence those who publicly denounce, and creates spectacles that rip people from place, territory, family, and home. Much like Taussig's "spaces of death" (1987), this mode of violence extends indefinitely for victims. It leaves no physical traces, no possibility of burial or conventional mourning rituals, no gradual organic decomposure that feeds back into a recycling continuum of life, which, of course, always includes death, dying, transformation, and change. The local fish thought to have fed on massacred bodies tossed over bridges are no human's future meal. "We are not eating one more damn fish from these rivers," I heard people say. The *limpieza social* (social cleansing) operations of paramilitaries share similar logics with state policies to eradicate illicit crops. Both rely on violent uprooting to pry open the space for what is considered to be a proper and dignified life.

Antonio's initial reflection speaks to these brute forms of violence. As he continues with the *mística*, he addresses the more subtle and insidious forms of rural dispossession and corporate chemical-seed complicity in these processes:

> We seek an involution into the land by way of seeds and not an evolutionary process where we leave them [the seeds] behind. One learns about chemical dosage from agronomy, pure mathematics. But they made us forget how to sow according to the cycles of the moon. We say that we are farmers, but this is a lie. We stopped being farmers once we started buying all our seeds, which they sell to us at any price. We now have to ask corporations for permission: What can we eat? What can we grow? And how can we grow it? These questions are determined by what we can sell because this is what they tell us that consumers, livestock, or cars eat.

"They" in this quote refers to a range of actors and institutions, such as state agricultural technical assistance, Colombian Agricultural and Livestock Institute (ICA) officials, USAID agricultural extensionists, and the random agronomists who visit rural communities to sell them commercial seeds and chemical inputs, implement official seed certification programs, inoculate crops, and advise on the implementation of dominant agrarian models. Commercial seeds and *criollo* seeds, or what are also heterogeneously referred to as traditional, ancestral, and autochthonous seeds, have long coexisted in tension. As Laura Gutiérrez's (2017) work on seed sovereignty struggles in Colombia points out, the persecution and/or denigration of criollo varieties of maize, bean, plantain, *guarapo*, *chirrinchi*, and other staple crops and foods—scornfully called Indian or poor people's food and crops—is a long-term historical process. However, many rural communities cite the ICA's passing of Resolution 970 in 2010 as the contemporary onslaught of the persecution of criollo seeds, and of the individuals found to have what are categorized in this resolution as "uncertified" or "unselected," and hence illegal seeds in their possession.[1] Much as Delgado and Rodríguez-Giralt (2014) point out in their discussion of the attempts of Brazilian campesino movements to include criollo seeds in Brazil's national agrarian insurance system, these popular seeds are contrasted with commercial ones based on their embodiment of diversity over generations through their cultivation and protection in situ by campesinos and landless and indigenous peoples. However, these longue durée processes always include endless adaptations to local agroecological conditions and attachments to complex colonial histories of displacement, deterritorializations, relocalizations, and attempts to remain in place.

I had been invited to accompany Heraldo and another friend from Mocoa, Carlos, to attend this 2010 seed exchange. We joined at least two hundred campesino, Afro-descendant, and indigenous peoples representing social organizations from around the subregion known as Alto Putumayo, the state of Nariño, and the neighboring country of Ecuador. The Network of the Guardians of the Seeds of Life had been founded eight years earlier by a group of campesinos and alternative agricultural practitioners outside Quito. The network became increasingly concerned with protecting criollo seeds from the homogenizing and privatizing tendencies of neoliberalized industrial agriculture. This particular workshop was held in the municipality of San Lorenzo, Nariño, in the vereda known as Valparaíso. Heraldo, Carlos, and I traveled for four hours with a group from Pasto through the Juanambú Canyon on an open-air *chiva* bus.[2] Upon arriving at the bamboo maloka, which was still under construction and had been designed to take the shape of a bird, we were greeted by Valparaíso and neighboring vereda residents, and colorful banners bearing the names of the community and regional organizations in attendance at the event: Agro-Environmental School Huellas, Campesino Mission, Our School, Community Organization of the Social Network of Families Las Gaviotas, Agro-Solidarity Federation, Committee for the Integration of the Colombian Macizo (CIMA-FUNDECIMA), and the Foundation for the Southwest and Colombian Macizo (FUNDESUMA). Vía Campesina and Latin American Coordination of Rural Organizations (CLOC) banners were also draped from the rafters of the maloka.

The vereda Valparaíso is located in what is known as the Macizo Colombiano (Colombian Massif), a nesting of mountains within the Andes that harbors an estimated 70 percent of the country's fresh and potable water. The Macizo is made up of thirteen *páramos*, 362 high mountain bodies of water, and the headwaters of the Magdalena and Cauca Rivers that flow northward to the Caribbean Sea, the Caquetá and Putumayo Rivers that flow into the Amazon Basin, and the Patía River that flows toward the Pacific Ocean. In lieu of playing the national anthem, which generally initiates government and community-led meetings alike, the seed exchange began with Antonio's mística and the sounding of the "Marcha del Macizo," a leftist protest song that describes the regional struggles and allied resistance of indigenous and campesino communities. All the workshop participants from Nariño rose to sing along, waving their *ruanas* stamped with the letters CIMA-FUNDECIMA.

"We came to realize that we had lost the mysticism and faith in our processes. We asked ourselves how it was that we began to detest and reject *lo nuestro* (what is ours/what comes from us). We have seen so much failure wrought by agricultural technical assistance that is not campesino-to-campesino or indigenous-to-indigenous based," Antonio says to the group. Eleven years earlier, the shifting local dynamics of war made it difficult for male community leaders to convoke meetings that exposed them to a dangerous level of public visibility after paramilitaries occupied and attempted to exert social control over the region. The women began to strategically gather in their gardens to continue to propel the community organizing work. Patricia and Elizabeth, two members of the directive committee of the Social Network of Families Las Gaviotas, explained that the women conceived of community mobilization as scaling up and spanning out from the level of their gardens, which they had previously called home gardens (*huertas caseras*). These home gardens gradually came to be designed and transformed into integral gardens (*huertas integrales*), and then collective spaces of learning and exchange or agro-environmental schools.[3] Natasha Myers asks the ethnographic question, "What is a given garden designed for?" (2017, 297). Among the families in San Lorenzo, they began to transform the design of their gardens to set broader political, economic, and socio-ecological transformations in motion. Twenty-six of San Lorenzo's fifty-four veredas now have campesino-to-campesino–taught agro-environmental schools led by what these women call *agrosembradores* (agro-sowing families). The schools do not rely on timelined project funding, but rather are in permanent processes of formation.

The agrosembradores recovered a conceptual practice utilized by their grandparents to distribute the daily labor and learning/teaching processes of the school: *hacer oficios* (to do the tasks or chores, to occupy oneself with the work). For example, some families are in charge of learning how to recover and transition soils away from chemical input substitution methods; others focus on learning water conservation strategies; others concentrate on protecting and circulating traditional and ancestral seeds; still others conduct research on and sow medicinal plants. Alba Portillo, the founder of the Network of Seed Guardians of Southern Colombia, explained that the protection of seeds entails producing and reproducing them through sowing and eating practices rather than simply cataloguing and conserving them in a community seed bank. Much as in the active circulation of criollo seeds, the methodology of the agro-environmental schools and integral gardens is to experiment and continually relay what one learns. These experiments seek to facilitate a gradual reconversion from agrochemical-dependent agricultural models that are productivist instead of production-oriented. The latter refers to the production of food and medicinal remedies that support the rebuilding of diversified local economies, the proliferation of agro-biodiversity, the strengthening of food autonomy, and the collective determination of rural communities. The municipality of San Lorenzo is unique in that several former agronomists who once sold agrichemicals in the area later returned to set up integral farms and support the community-led agro-environmental schools. This occurred after they were inspired to unlearn agronomy through different processes that they described to me under the umbrella of "becoming agroecologists." The next two days of the workshop in Valparaíso were spent rotating among smaller groups to learn how to catch rainwater and compost human feces, to share community experiences establishing seed networks, and to exchange gardening and reforesting practices.

The reclaiming of seed diversity attempts to address what these rural communities diagnose as their forced amnesia. Amnesia is perceived to be a result of the marginalization and destruction of popular, traditional, and ancestral agricultural practices as well as their underpinning ethics and economic structures by the colonizing and modernizing forces of capitalist agriculture. The destruction of practices has been paired with a systematic lack of opportunities to learn from alternative forms of agricultural techno-scientific assistance, such as agroecology and agroforestry, among other less consolidated or recognizable experimental practices that emerge among exchanges (*intercambios*) between ethnically diverse rural communities.

At the workshop in Valparaíso, people began the conversation by listing the varieties of seeds that had been "lost" in their home locales. In Mocoa, Putumayo, this included the sweet orange trees that commonly grew in people's yards, *yota* (an Amazonian substitute for the Andean potato), *maíz indio*, *maíz pira*, and *puntilla* corns, countless varieties of beans, *pan de norte* that makes for a delicious alternative to corn-based *arepa*, and Amazonian cacao, which is smaller but more robust than industrial cloned varieties. There had been at least sixty identifiable varieties of palm trees with edible fruits, oils, and leaves, yet people now only consume or transform products from twenty of these palms. In San Lorenzo, Nariño, over fifty creole varieties of coffee, sugarcane, and corn seeds had been reduced to two or three commercial varieties. In Ecuador, the seeds had different names and uses, but the situation was parallel. People also referenced protective or "magic" plants that defend the forest, harmonize, and bring good luck and prosperity, and that had all but disappeared from farms and gardens, such as *ortiga* (nettle) and the *caspi* tree, or what others call the Pedro Hernández tree that provokes welts and fevers in ill-intentioned and disrespectful people who do not ask for permission when in its presence. Not far from Pedro Hernández always grows its curative counterpart, the *espadero* tree. Heraldo provoked empathetic laughter among the group when he shared how he had planted *tigre* (tiger), *borrachero* (drunkenness), and *soplón* (snitch or whistleblower) plants at the entrance to his farm to ward off thieves who creep up to steal hens or tools in the middle of the night, and those that frequent the farm in broad daylight fishing for votes during election season—what he calls *liberales y godos*, the liberal and conservative politicians who had historically dominated regional elections in Putumayo. Inspired by this creative play on words, I began to call these plants *espantapolíticos*, which is a play on the word for scarecrow to make "scare off politicians."

At one point in the conversation, Heraldo leaned in and whispered to me that criollo seeds are not necessarily on the road to literal extinction, although in some cases this is also true, but more accurately, many seeds had gone extinct from people's minds. They no longer had a mental inventory available to them. They had become strangers with no affective ties or social relations, similar to Vandana Shiva's (1993) now well-cited description of "monocultures of the mind." To name a few examples, he cited *diente de león* (dandelion root), cilantro cimarrón ("wild" cilantro), *mafafa de hoja* (mafafa leaf), *espinaca de bejuco* (creeping plant spinach), *ajo de monte* (forest garlic), and *maní estrella* (star nut). These seeds, he said, needed to be "resocialized" in gardens and orchards, across kitchen tables, on the taste buds of tongues, and moving through human and animal intestines. Rural communities should ask themselves how they had come to live in a world where "wild" plants and "nonresources" were devalued in comparison to intentionally cultivated and commercial plants, he told me. "Why do people want to eat imported apples, pears, and mangos when we have Amazonian fruits, such as guamo [a leguminous tree that produces a large pod with a sweet pulp], *caimarón* grapes, *zapote*, and forest guavas?"

Yet Antonio's opening *mística* also inspired me to reflect on how even when rural communities have grown unacquainted with or been taught to repudiate many varieties of criollo seeds, these seeds persist by becoming imperceptible—that is, by receding into the background and converting into inconspicuous "weeds" along the sides of highways and roads, or retreating into gnarly vegetal masses of what, for many, was nothing more than indiscernible monte. Of course, criollo seeds also continue to be quietly sown in gardens and ancestral chagras by humans who stubbornly and caringly conserve their sociality. They are also transported by animals through their digestive tracts and defecation and are dispersed by the air currents of wind and breeze. Far from *maleza* (*mal* meaning bad and maleza being weeds), Heraldo told me that most of these plants are *bueneza* (buena as in good) because they provide medicinal remedies, associative nutrient cycling, soil cover, and scents and vibrations that attract and repel different insects, animals, and peoples.

The colorful fistfuls, paper bags, and satchels of seeds that were later exchanged at the workshop reminded me of the Catholic priest, Padre Alcides Jiménez. Alcides was well known for distributing traditional Amazonian seeds instead of the host during the masses he led at the Nuestra Señora de Carmen church in the municipality of Puerto Caicedo, Putumayo. People remember that the pockets of his cassock were teeming with seeds, and that he reforested the parish patio until selva was practically bursting through the windows of the vestry. I unfortunately never had the opportunity to meet Padre Alcides, but I did go to the chapel where FARC guerrillas are said to have assassinated him on September 11, 1998. A small room at the front of the church preserves the bullet-riddled Bible he was holding on the day of his murder; the punctured chalice that somehow remained steady and avoided spilling wine on the altar cloth; the pair of reading glasses he toyed with nervously during the sermon when he recognized that hooded assassins were lurking in the nave. In 2009, I collaborated on the curatorial work and translation of an itinerant memory gallery commemorating victims of violence that was supported by MOVICE (the National Movement of Victims of State Crimes) in Bajo Putumayo. This project allowed me the privilege of meeting Padre Alcides's brother and learning about the twenty years the priest spent accompanying rural communities in Putumayo.

Padre Alcides established a community outreach program called Corporación Nuevo Milenio. Among other activities, the program promoted polyculture, Amazonian-based farming—one aspect of his ecological evangelism that preached against monocrop agriculture, coca and otherwise. Alcides equally criticized the repressive and indiscriminate nature of the state's aerial fumigation policy and the economic dependencies and environmental degradation produced by coca-growing rural life-worlds that no longer sow a diversity of seeds from which to produce their own food and remedies. From his pulpit, he opposed the 1998 armed strike endorsed by the FARC-EP because of the imminent paramilitary retaliation it provoked against civic leaders and human rights defenders. Instead, he proposed a position of "active neutrality" toward all the armed actors occupying the territory. Alcides promoted a humanitarian dialogue based on the dictates of international humanitarian law and armed actors' agreement to respect community autonomy, including community efforts to transition out of illicit coca production. In a letter he wrote two years before his death, Padre Alcides celebrated the existence of gardens with over sixty varieties of edible and medicinal plants in the midst of what was otherwise an almost monopolized reliance on commercial coca crops (*Semillas de Paz* 1996). I wondered how he would have felt upon visiting Heraldo's two-hectare farm, where we once conducted a rudimentary inventory and counted over one hundred varieties of plants and trees, which did not include the many plantain and cacao trees and miscellaneous shrubs that Heraldo was unable to identify by taxonomic or popular names.

What I refer to as reclaiming and relaying marginalized and destroyed practices is not a romantic quest to rescue an authentic past. There is no triumphant heroism in the act of reviving solar/lunar rhythms in agricultural cycles, applying an avocado seed in place of chemical insecticides, or circulating and protecting uncertified seeds. Selva that reclaims a church vestry and hands that pass out seeds instead of holy bread are riddled with bullets. Rober Elio, a campesino leader I met and befriended at the seed exchange in San Lorenzo, later lost an eye when the Mobile Anti-Riot Squad (ESMAD) shot tear gas into the faces of protesters during the 2013 National Agrarian, Ethnic, and Popular Strike. Rural community organizers have been systematically imprisoned on false charges of rebellion, and uncertified seeds are criminalized and confiscated by the state in defense of the intellectual property rights of multinational corporations, such as DuPont, Syngenta, and Monsanto-Bayer, in accordance

FIGURE 4.1 Don Nelso's hand holding bean seeds in the process of cleaning them.
Photograph by author.

with the dictates of the free trade agreement that Colombia signed with the
United States.[4]

I borrow the term *relaying* from Isabelle Stengers (2017) to highlight the
ways in which the reclaiming of practices that have been destroyed, or were
intended to be destroyed, is necessarily an active *adding to* and not simply a
critical denouncing of. Relaying demands consenting to an ongoing process in
which the relayer "casts [their] lot for some ways of going on and not others"
(396)—that is, for some ways of living and dying together, sowing and eating,
shitting and decomposing, working and exchanging, organizing and transforming, and not others. Relaying and casting, of course, involve differential
political stakes, vulnerabilities, and corporeal exposures. There is no outside or
self/collective assured position from which to take action. There is no guaranteed paradigm to follow when rural families engage in their everyday work to,
as I earlier explained in Heraldo's words, "technically emancipate a territory,"
or what some families I accompanied described in terms of "decolonizing their
farms." Inspired by what I came to think of as *pulsations of cosustainability*—
dispersed selva practices striving to resonate and multiply across the Andean-
Amazonian foothills and plains outside existing modes of agrarian regulation,

I spend the rest of this chapter engaged in an ethnographic focus on two families who form part of the alternative agricultural network influenced by Padre Alcides's ecological evangelism and the legacy of his community organizing.

These families shared with me how they are learning from the recycling conditions of hojarasca—rot, decay, and regeneration, and the ways that this de- and recomposition becomes integral to the cultivation of experimental practices among farms, gardens, and forests in the midst of war and continued socio-environmental conflicts. What most compelled me is how the practices attached to these processes of decomposition enable not merely people, but entire ecologies to strive to transform the conditions of their shared existence— not by transcending these conditions, but rather by sinking into them, slowly turning them over, aerating, breathing in new life that also potentiates different possibilities for and relations with death. These rural families taught me that a continuum of organisms and elements might resist violent modes of death by becoming into death instead of working against it in the pursuit of a "better" life. By becoming into death, I refer to a mode of dying that is an aspect of the transformation of being, an emerging into many other living and dying things much like the regenerative decay of decomposing leaves. This mode of death results from the successional recycling of selva, which is quite different from the violence of war that brutally severs people from land and territory—or, as I discussed in the previous chapter, a war that goes by another name and, what is the same, kills in the defense of extractive-based and unlimited economic growth.

SUCCESSIONAL EXPERIMENTS:
ROOTLESS PROPOSALS NEVER GROW IN ONE'S HEART

Micos boca de leche (milk-mouthed monkeys) rustle in the highest branches of the guamo trees that provide shade for the farmhouse. They draw my attention upward as we walk toward the entrance of the dense successional agroforestry experiments that Álvaro and his wife, Marta, began cultivating seven years ago. Their neighbors claimed that it was ludicrous to sow all their seeds at once, convinced that nothing would come of what was sure to be an ensuing tangled mass of vegetation. Along our way, we pass three hectares of neatly regimented monoculture heart of palm trees. These trees were part of a USAID alternative development project linked to the heart of palm factory in Putumayo's commercial capital, Puerto Asís. The factory managers had recently decided that it was no longer profitable to travel to Álvaro and Marta's vereda to collect the harvests. Heart of palm was now rotting everywhere in sight. Marta could not serve it up fast enough: *palmito* for the breakfast, lunch, and dinner of the dogs, cows, human family, and visiting anthropologist. The USAID agronomists hired to provide technical assistance for the project convinced the family that heart of palm needed to be sown as a monocrop with total exposure to sunlight. They arrived at the farm with what Marta and Álvaro characterize as a "take it or leave it attitude" and start-up supplies: fifteen sacks of chemical fertilizer, twenty-nine sacks of lime to "correct" the acidity of the soil, nine sacks of organic inputs, some local varieties of seeds, and seeds that were introduced from somewhere else, but that were never clearly explained to the family. Álvaro tells me that since a monoculture system is one in which all the plant rootlets compete for the same macro- and micronutrients, each hectare of palmito continually requires four tons of organic inputs. The USAID project was fundamentally unsustainable. When technicians from the municipal agricultural extension office (UMATA) later visited the farm and affirmed that heart of palm trees would not tolerate any shade, Álvaro and Marta told them, "You have a theory and we have the practice. We are interested in a vision of production that follows the selva and allows her to complete her cycles by minimizing the damage to them." The family admitted that they had been duped by the agronomists and were now living in the aftermath of the destruction of organic material that they themselves had provoked. Don Liberman, a neighbor who arrived to join the day's minga (communal labor) added, "It is ironic to work with plants that we don't eat. We don't understand their behavior. We have no relations with them. Rootless proposals never grow in the hearts of campesinos."

Before Álvaro begins to explain to me the processes of natural and assisted regeneration of forest that are occurring on the farm, we are reminded to keep our senses attuned, rotating from trees to ground, shrubs, puddles, creeks, mud, manure, and grass. The family dog grows increasingly excited and teasingly paws at something in the high grass to our right. We hear only the faintest whisper. Blades of grass suddenly splay open, are quickly pressed down, and rolled over. A thick and shiny body glides by. I catch just a glance of its back and tail, but a tremendous length and weight is implied. The dog suddenly lets out a high yelp and comes running toward us. His snout is bloody and he whimpers at our feet. He has suffered a potentially deadly snakebite, and by the afternoon, pus-filled wounds have opened up all over his body. There is not much to do other than to comfort him and provide water. Álvaro calls the uncle of one of his son's friends, and the man performs a "secret" that either will kill the dog if he is in too much pain or will choose to save him. The next morning, we have our answer. The dog's snout has begun to scab, and he moves slowly, with disorientation. The gradually waxing light in his eyes reflects that the "secret" elected healing.

Álvaro and Marta do not kill the many creatures that visit and inhabit, and thus make their farm, without much deliberation. They attempt to teach their grandson, a young boy who enjoys squishing bugs, the same. They do not kill venomous snakes. Some eat the mice that feed off of their *chontaduro*. They do not kill the venomous conga ants. They do not dismantle the nests of bees and hornets burrowed in the lemon trees. They do not rid themselves of the foxes that occasionally eat the hens. They spray their crops with wild garlic, onion, *barbasco*, and chili pepper juices because their intention is not to eliminate "pests," but, as they explain, to "urge certain creatures to be on their way." Often this means straight into the beaks of avid birds hovering above a future meal. They apply mineral broths (*caldos minerales*) made of ground meat bones rich in phosphorus to strengthen the plants against bacteria and fungus. They employ preventive and not curative strategies, and this entails the human family, along with myriad other creatures, to remain continuously attentive to the comings and goings, seasonal cycles, shifting climates, hiving, appetites, metabolisms, pollination patterns, and attraction and repulsion of different bodies and elements. The only creatures that Álvaro and Marta consider definitively threatening are the *temblones* (electric eels) that reside in the dark waters of the creeks they have reforested as well as the large puddles that crisscross their farm. If children or animals fall in, they will likely be electrocuted or pass the electric shock on to someone else after climbing out of the water. They are not interested in eliminating all the eels, but only in reducing the population. As they understand it, no one has the right to simply invade the home of another. Neither is it acceptable to invite creatures into their home while imposing how, where, and when they can move, eat, sleep, shit, pollinate, and urinate.

Don Álvaro, who was fifty-two years old at the time of my fieldwork, migrated to Puerto Caicedo from Nariño with his parents when he was twelve years old, and he is the only member of his family to remain in Putumayo. A year and half earlier, when he traveled to the city of Palmira, Valle de Cauca, to visit his ailing mother, he himself grew ill from the claustrophobia of being enclosed within four walls. It was the loneliness and cagedness of the city that haunted him. "One comes down out of their apartment to buy food at a corner store only to turn around, go up, and climb back in the box. Here in la selva it is not just the human neighbors who recognize me, but also the plants, animals, and trees," he told me. This selva sociality is a contingent and precarious affair because of potentially venomous encounters with snakes, electric eels, spiders, and ants, and also due to the shifting conditions of war and the perpetual threat of violent displacement and dispossession. Between 2000

FIGURE 4.2 Don Álvaro explains the life philosophies of his family's successional agroforestry experiment. Puerto Caicedo, Putumayo. Photograph by author.

and 2003, FARC-EP guerrillas arrived at the family's house demanding to know why Álvaro refused to attend their obligatory community meetings or arrive at the hours and locations they stipulated. Then paramilitaries visited the family demanding to know why the FARC had been frequenting the farm. Soon after, the paramilitaries built a small detention center on Álvaro and Marta's property that was located two meters from their front door. No one was allowed to fish, walk freely in the forest, be outside after six p.m., or milk cows early in the morning for the daily pickup by an intermediary who sells the milk in town, which is one of the family's modes of economic sustenance. The paramilitaries murdered Álvaro and Marta's nephew, and Marta's nerves collapsed. She left for two years with their young daughter and son while Álvaro stayed on to work the farm as best he could, living in fear that paralegal armed actors would appropriate more of their land. Over the previous eight years the spray drift resulting from aerial fumigation operations doused their farm with glyphosate at least four times. As soon as they heard the whirl of the crop-duster planes they would run outside to save the plantains and cover their water sources. The earth had aged and grown tired. Strange weeds had begun to grow. The insects were not the same. Many things were amiss. Marta told me, "if these herbicides can burn through sticks and stalks, which are much more resistant than a human body, imagine how they must be affecting us."

Álvaro and Marta's twenty-eight-hectare farm, which came to include part of Marta's elderly parent's property, is situated in the fertile floodplains or *suelos de vega* of the Putumayo River. The river is located fifteen minutes away and the mineral-rich sediment deposited by its recurrent deluges nourish the area every six years or so. Ground-level gardens and crops are inundated and washed away so that future gardens and crops can grow. Fecundity is contingent upon cyclical destruction, and these sequences of decomposition and renewal resonate with successional agroforestry logics. Life on the farm begins at 3:45 in the morning when Álvaro and his son milk the cows for the five a.m. milk collection. This is a seven-days-a-week job. They then eat breakfast and rest for two hours until the day begins for the second time. They labor on until noon, when the heat can become overwhelming. On other days they work straight through until five p.m. When downpours occur, it becomes a day of rest with a double workday to follow. The family tends to small-scale dairy cattle, chickens, and hens, and there is corn, copoazú, sachainche, and arazá to harvest. Half of the farm remains primary forest and the other half is divided into plantain and yucca cultivations, the failed USAID heart of palm project, *guadua* (bamboo) trees that can be sold as building material or for fencing and artisanal furniture, and the two experimental successional agroforestry gardens that I have come to visit. Álvaro explains to me that it is impossible to live from a singularizing agricultural model because this generates excessive market dependency. One can lose a crop or, as in the case of the palmito, one can be excluded from a particular niche market, or prices can abruptly fluctuate. The family practices a diversified economy that depends on little predictable relationship between economic activities and singularizing vocations, and thus they avoid being categorized according to their economic involvement as *bananeros* (banana growers), *lecheros* (milk producers), and so forth.

Álvaro, Liberman, and another neighbor who has arrived, Alquimides,

FIGURE 4.3 Maturing sachainche, otherwise known as maní estrella (star nut). Puerto Caicedo, Putumayo. Photograph by author.

explain the philosophies attached to their successional agroforestry. "The idea is not about making money, but about making way for life. Amazonian agriculture is polyculture, successional, varied, and collective modes of production that help the family economy and the ecology." The crops go on varying, but within the same space. This is a space that farmers have "freed up" from the practice of beating back the invasive grasses that grow in deforested fields while simultaneously safeguarding or, as they call it, "freeing up" nearby forests from future predation. For example, in one of Álvaro's successional arrangements the bean crop reaches maturity after 120 days, the corn between

120 and 140 days, the Amazonian wheat at 150, the *flor de Jamaica* (hibiscus flower) between 150 and 180, the yucca between five and six months, and the plantain trees once a year. Nitrogen-fixing leguminous varieties of trees, such as *bilibil*, guamo, and *macuna* are dispersed between the plants along with small *achote* shrubs whose seeds are used to color and flavor food, and timber-yielding trees are sown ten to twelve meters apart. Álvaro tells me, "I have a soup of plants growing here." All the seeds are sown simultaneously, but what are referred to as pioneer, primary, and secondary plants grow successively. One never clears the vegetation; rather, it is trimmed back to create a space for each plant. I would often hear Heraldo refer to this as the creation of a "vital space." I expand upon this concept in the next chapter.

These campesinos tell me that asking "Can I grow this particular plant?" is entirely the wrong question. If a plant cannot be eaten or sold by the human family, other creatures or the selva will eat it and shit it out, feeding on and decomposing back into hojarasca that is always already becoming nutrients, air, precipitation, food, shit, becoming hojarasca. Álvaro explains that he is particularly attentive to the vengeful nature of the selva and the need to follow and not eat against her. These men patiently explained to me the lessons they are actively learning, such as becoming attentive to the differing heights and strata of each plant, their convergent and divergent needs for light and shade, and the durations of their life cycles. The ability to select a principal cultivation and associate it well within a successional continuum is especially important given that every plant requires more or less work from a family and produces differing amounts of organic material and shifts in the emerging agroforest's microclimates, which attract and deter different insects, bees, butterflies, and microbial communities. In part, this is why it is crucial to have a diversity of criollo seeds available, which is one of the greatest challenges that most of the rural communities I met in Putumayo face, given the relative homogenization of seeds produced by the decades of monoculture coca crops and their official alternatives.

For example, much like Heraldo, Álvaro and Marta plant Amazonian wheat rather than commercial Andean varieties. The former is used to feed hens and to avoid buying commercial feed and potentially genetically modified corn. In place of growing a temperate commercial fruit known as *lulo*, they grow the smaller Amazonian variety that is locally consumed called *cocona*. They plant sachainche and *inche* Amazonian nut trees that most of their neighbors fail to recognize and thus cut down. As we walk toward the second and more recent successional agroforestry experiment, Álvaro bends down to dig up a *yame* (Amazonian potato). He tells me there are at least three varieties,

FIGURE 4.4 *Trigo amazónico* (Amazonian wheat). This grain is used to replace commercial concentrate to feed creole hens. Photograph by author.

and that the most delicious one does not yield fruit as often as the others. Less is more. The neighbors have gathered today to prepare this quarter hectare of land, which is in the process of becoming a successional agroforestry garden. The pioneer plant that Álvaro has chosen is an Amazonian cacao, copoazú, rather than plantains as in the previous plot. Within a week, copoazú seedlings will be sown, leaving four meters between plants and five meters between rows. Arazá and cocona Amazonian fruit trees will be interdispersed at two to four meters distance, and timber-yielding varieties at ten meters. The corn will produce three to four harvests a year and the kernels will be scattered directly amid the hojarasca along with the Amazonian wheat, kidney and habichuela beans, and sorghum seeds. Some trees have been cut down and they are now being chopped and minced to create more biomass to enrich the ground. These campesinos tell me that they need to keep chopping and mincing as if preparing a salad, so that eating, metabolizing, defecating, decomposing, and germination can carry on.

As I outlined in the first chapter, protracted neocolonial histories of extractivism entangled with multiple waves of displacement, territorial appropriations, and settlement severely and differentially impacted the capacity of rural inhabitants, especially indigenous communities, to engage in regenerative forest processes. These regenerative processes traditionally depended on their capacity to circulate and cultivate across large tracts of ancestral territory that could then be left to rest and recover over extended periods of time. All three of the men I met that day and their families had participated in Padre Alcides's agroforestry and integral Amazonian agricultural projects. Latin American and European agroecological scientists, alternative technicians, and environmentalist donors had at times visited these projects and temporarily accompanied the farm school experiments with local communities. The assassination of Padre Alcides and the subsequent threats made against the organizers and members of Corporación Nuevo Milenio by paramilitaries caused community efforts to recover successional agroforestry practices, criollo seeds, and medical plants to become more dispersed and go quasi-underground. As I have mentioned elsewhere in the book, Amazonian or selva agricultural practices are not widespread beyond a scattered network of collective initiatives, rural families, and individuals. Yet they form part of the ongoing twenty-year-plus political struggle between the state and regional social movements, as the latter oppose official antidrug policy and extractive-based economic development models. Marta and Álvaro told me that they felt fairly isolated and worried that alternative and ancestral agricultural practices were not being fostered and shared. While we spoke, Marta offered me a sample of flor de Jamaica wine, and excitedly told me that there is now a waiting list to buy it. She is teaching her neighbors how to make it along with other recipes inspired by the successional garden, such as copoazú cakes and cookies.

In many ways, the family's activities parallel those of another family that I became extremely close to and who had also worked with Padre Alcides during the 1980s and '90s. Don Nelso and Doña María Elva have recovered and creatively innovated the preparation of more than three hundred Amazonian recipes. They established a multifaceted family business called Productos Amazonia to both sell these foodstuffs and teach others, especially

Mocoa's network of displaced women, how to prepare recipes from the diversity of native and adapted vegetation making up integral Amazonian gardens and orchards. One example of a recipe is what they refer to as *copulate*, which is a play on chocolate. It is made from Amazonian cacao, copoazú, and not the cloned industrial varieties. Another of their specialties is *limonayo*, which is a play on *limonada* (lemonade) and the traditional Aruahaca word for coca, *hayo*, or a combination of lemonade and ground coca leaves. Elva also experiments with baking breads, cakes, and cookies from fruit of palm, plantain, and wild ginger flours as an alternative to commercially purchased enriched wheat flour.

CREATING CONDITIONS AND LIVING EN MEDIO DEL RASTROJO

The story of Nelso and Elva's forced return to Mocoa pivoted on a kind of negotiated displacement between Nelso and members of the paramilitary group AUC after they occupied the town of Puerto Caicedo. The paramilitaries explained that they had no real reason to chase Nelso out of town, as they had been watching him for months and found no evidence that he was collaborating with FARC-EP guerrillas. However, other people wanted him gone, and as is often the case in contexts of war, these families made use of a generalized dynamic of violent displacement to wage personal vendettas. Nelso acquired enemies as a community leader working with Padre Alcides's programs to encourage coca growers to stop cultivating the crop before the state's threat of aerial fumigation homed in on the region. He also promoted a local river clean-up and restoration project and was advocating for the mayor to build a separate market for small-scale producers, which would not only function for commodity exchange, but would also be a space for bartering practices, seed exchanges, cooking classes, and the gatherings of rural communities. This was in 2001. Visible seesaw trajectories between town and country drew immediate suspicion due to the strict territorial divide imposed by the FARC and the AUC. When I was alone in the kitchen with Doña Elva, she told me that Nelso was persecuted after he read a commentary on his weekly radio show, *Para el Campo desde el Campo* (For the Countryside, From the Countryside). Nelso publicly criticized the mayor for only paying attention to rural communities during election season. The community radio station immediately aired a disclaimer, running it continuously for the next forty-eight hours: "This does not reflect the view of the organization, but the opinion of an individual . . . opinion . . . individual." Elva begged them to stop the sound bite, but it was already the beginning of the end. The death threat that the AUC nailed to Nelso's front door read: "WE WILL MOW YOU DOWN WHILE YOU WALK HOME!" It was a quick decision. Nelso and Elva packed up a few belongings and headed north with their five children until they reached Putumayo's capital city of Mocoa.

What followed was a year, as they described it, of *quedarnos quietos* (lying low, staying still), a forced retreat from public life and the exposure inherent in too many social relations. They were lucky enough to receive a government-funded housing subsidy for displaced families, but when the first opportunity arose, they traded the one-room apartment for half a hectare of land on the outskirts of Mocoa. They had only recently obtained a formal land title, and for a long time a handwritten note served as a deed. "We were getting sick living in a matchbox," Elva tells me. She is bent over a cafeteria tray full of seeds, varieties of beans we inspect for tiny holes, a literal worming-in of uninvited mouths. These are "beans for eating, not for commercial purposes," Nelso explains, but of course, not all mouths are the same, nor are they fed equally. These are beans to feed other beans, beans to feed microorganisms and cloak naked soils, to gift and trade; they are also for the family's lunch and dinner—red, black, speckled, gray, maroon, shades of café and mustard. Not only seeds, then, but mouths as well. One by one these seed-mouths are turned over in the palms of hands and stored again, a laborious practice to cleanse rather than submerge them in chemical insecticides. Seed storing practices always depend on particular regional climatic conditions, such as levels of yearly rainfall and humidity, which create a propensity for bacteria and fungi to more speedily proliferate in the Amazon. Elva explains to me how the family stores their seeds in glass jars, covered in ash, and sometimes buried in drawers between pieces of clothing.

Glancing over at their neighbor's plot, I begin to envision what the area looked like more than ten years ago when they first arrived. A stretch of rust-colored earth, compacted by cattle hooves, burned out by herbicides, open-faced, and exhausted. "Seemingly left for dead," says Nelso, referring to a state of being which, as I explain later, can be conceived as refusal—a state of imperceptible playing-dead through which regenerative transformation can be achieved. In the case of their neighbors, like so many farmers following the Green Revolution technical advice of agronomists advocating for the use of chemical inputs—and to whom Nelso refers as *ladrónomos* (or thiefgronomists)—the "natural limitations" of Amazonian soils have to be continuously struggled against and corrected. His punning neologism for the agronomists speaks to the softer kinds of violence exerted through dominant techno-scientific paradigms. To combat the "poverty" of the soils—their thinness, acidity, old age, and mineral deficiency that I discussed in depth in the previous chapter— Nelso describes the conventionally prescribed technical treatment plan: drips

FIGURE 4.5 Doña María Elva converses with me about Amazonian seed storing and sowing practices. Mocoa, Putumayo, January 2013. Photograph by author.

of urea and Triple 15, a chemical pump strapped to the back of a farmer, the hissing sound it makes close to an ear like an old man's raspy breath. An entire life support system is assembled not only to resuscitate the soil, but to force its betterment. This is not engagement with a body that is recognized as having corporeal finitude and limitations, but rather the treatment granted to artificial strata whose mortality is not only cast as a weakness to be overcome but is all but denied. Of course, there are other methods that farmers use to render post-coca fields productive, such as planting monoculture pineapple, sugarcane, and other acid-tolerant crops. However, this seemed to Nelso and Elva more of an economic contingency plan and a denial of fragility and tenuousness as intrinsic to the ebb and flow of life.

In stark contrast to the chemical optimization of the soil's regenerative capacities, Elva spoke about "creating conditions." Conditions that do not forcibly supersede corporeal vulnerability to resuscitate soils but that allow for the kind of withdrawal that the family had seen its own life cyclically retreat into and slowly emerging back from: underemployment, displacement, landlessness, lying low, and the ongoingness of recovery. They were not in the business of imposing conditions to make productive soils or—in a broader sense—

FIGURE 4.6 Sorting seeds with Nelso and Elva. We counted thirty varieties of beans in their integral Amazonian garden. Mocoa, Putumayo, December 2010. Photograph by author.

productive lives emerge. Each day Nelso and Elva bring together seeds, manure, wood chips, eggshells, husks, and vegetable and fruit skins to create conditions for their intermingling in anticipation of what might emerge. One example is the way decomposing elements—litter layer, ash, rinds, rootlets, manure, human urine, sugarcane mash—come to mutually constitute each other as they transform into hojarasca, which holds their original traces while also constituting something new that is always already changing. They initially collected cow and horse manure from farmers who did not incorporate it into their fields and made daily trips to Mocoa's central market to fill a cart with rotting organic matter. They did not own a horse, and these trips went on until Nelso's back gave out, and both Elva and his aching vertebrae convinced him to stop lugging loads of vegetable scrap up the hill to their farm.

They began by collecting the cans, tires, and plastic bags left behind by the previous residents. They discussed which plants would be their allies in interrupting the life cycles of the domineering pasture grasses that flourish in open sunlight. They planted sunflowers, low-growing kisses, yams, beans, cowitch, *cudzú* shrubs, and wild clover: food and protective covering for others, mostly

FIGURE 4.7 Living in the midst of animal fodder. Mocoa, Putumayo.
Photograph by author.

microbes that would nurture plant rootlets that would eventually come to
shade the soil. Then they waited, thinking, searching for building materials
to "create conditions" for the house itself. In this way their shared lives—soils,
farmers, plants, worms, microbes—emerged out of a retreat as necessary as it
was imposed. Many neighbors and agricultural extension experts looked on in
bewilderment as they watched them sow what appeared to them to be roadside
weeds. "*Los locos viviendo en medio del rastrojo*," they said—crazy people living in the midst of animal fodder, or selva that returns when left unattended
and that most farmers clear in the act of occupying, *mejorar* (bettering), and
achieving ownership over a plot of land.

Only after knotty bundles of diverse roots and organisms had, as Elva explained, "created a climate" was it acceptable for human hands to loosen up
and turn the insides of soils out, allowing for the aeration that invites life and
death—digestion, defecation, and decomposition—to proliferate. This was
an altogether different propelling of "insides out" than what paramilitaries
threatened to do to the bodies of *sapos* (whistleblowers) when they used a similar phrase to refer to the torturous interrogation techniques they employed to

force confessions. The "creation of a climate" could be partially established through what Nelso called a process of "getting to know a plant," similar to what I previously described as practices of lecturaleza. For instance, becoming attentive to the size of leaves, diameters of stalks, plant coloring, the timing of fruition, the presence of earthworms, and if flowers or buds had fallen off the trees prematurely. During one of our first conversations, I mistakenly tried to translate "getting to know a plant" into the scientific assessment terminology of bioindicators. However, this translation failed to work for the family, not merely because, as Nelso said, "they did not have a name for it," but also because processes of getting to know a plant always exceed what the plant reveals to a farmer, and thus what the farmer can claim to know of the inner secrets and worlds of a plant.[5] As Elva explained it, campesinos work for ants and microorganisms when they become aware of the ways that ants and microorganisms "create conditions" for all kinds of beings that far exceed the results of any human labor on the farm. She described feeling inklings and intuitions provoked by the "life force in everything" that orients her to engage in this or that activity in the garden. There was a mística to these encounters, and Nelso told me that when he had been away from the farm for several days he would sometimes hear the plants calling out to him: "¿Quiubo? ¿Quiubo?" (What's happening? How's it going?), they asked. He could not always hear them, nor should they be expected to say too much. However, on certain occasions he was able to reply and ask them what had happened in his absence. Such and such a thing had occurred—pesky critters, invasive grasses, or a lack of rain. These moments of lecturaleza orient Nelso on how to responsively go on "creating conditions" by inviting certain insects or plants to leave and others to visit the farm.

Later on, the family began to plant human food crops, mostly noncommercial (or at least not traditionally commercial) perennial varieties: for example, Amazonian wheat, medicinal plants, star nut, plantains, tubers, fruits, and greens for the hens, ducks, rabbits, and guinea pigs that provide manure for the farm and that are also easily stolen by petty thieves. Nelso and Elva's children did not show initial interest in farming, and instead took minimumwage jobs in town to support their own growing nuclear families. However, they continued to share the same house and eat from the gardens. Over time, they watched the farm convert into a dynamic space frequented by other families interested in trading seeds, produce, and recipes, and also town folks and restaurant owners who come to buy fruits, pulps, vegetables, nuts, and medicinal plants unavailable at the commercial markets in Mocoa. The commercial markets are stocked with produce and foodstuffs transported from the neighboring Andean states of Nariño and Huila. Shopkeepers pay a fee to utilize and maintain the vendor stalls while small farmers sit outside unprotected from the sun and rain to sell staple foods, such as yucca and plantains, or Amazonian grapes, white sugar pineapples, and other seasonal crops.

For Nelso and Elva and the other campesino families I accompanied, the acidity of regional soils—with a pH of 5.6 to 5.8 when soils in temperate zones tend toward neutral or 7.0—is not perceived to be a problem that requires correcting. On the contrary, it is a capacious characteristic that retains the nutrients that reach the ground through rainfall, but that would otherwise be quickly washed away by the region's heavy tropical rains. These nutrients are later solubilized by microorganisms and become available to feed plant rootlets. Lime may be added to the soil as a nutrient, but not as a corrective measure for acidity. The family does not engage in composting and argues that most campesinos find this to be exorbitantly labor-intensive. Instead, to nurture plantain trees, for example, they dig a meter-wide hole in preparation for where they plan to sow the tree and fill it with organic scraps that decompositionally enrich the area where the tree rootlets will later extend. Daily feeding of plants and trees occurs with care not to pile decaying organic material too close to the base of stalks to avoid the proliferation of bacteria and fungi. Like Heraldo, they sow their seeds at greater distances instead of planting them at the congested distances recommended by most agronomists. This is because they have learned that certain rootlets are able to vertically penetrate the acidic layers of clay to seek nutrients, such as sunflowers and sugarcane, while other rootlets splay horizontally, seeking nutrients in the shallow five- to ten-centimeter layer of hojarasca. Sunflowers are an excellent *planta recuperadora* (recovery plant) to associate with horizontally feeding plants because their leaves fall to the ground full of the metabolized minerals that are not accessible to the roots of their horizontally splaying neighbors. The family does not necessarily conceive of their farm as a successional agroforestry experiment. I prefer to think of it as they occasionally refer to it: an experiment in living in the midst of "animal fodder."[6] Nelso, Elva, Marta, and Álvaro, among other families, taught me that in order to "decolonize their farms" they first have to decolonize dominant techno-scientific, chemically conceived, and market-oriented notions of and relations to soils, plants, trees, and seeds.

THE CULTIVATION OF COUNTERLIFE AND DEATH

The different waves of structural and armed violence in the late nineteenth and twentieth centuries that expelled families from the country's conventionally fertile Andean region into frontier territories impacted by "poor soils," and the internal dynamics of war and displacement in these territories, necessarily transform people's attachments to life. The diverse families, campesino associations, and alternative agricultural practitioners I met did not tell me stories of social suffering and state abandonment or endurance against all odds of precarious life (Biehl 2005; Nouvet 2014). Nor did they dwell on modern life rendered meaningful only by denouncing its finitude, and thus animating biopolitical logics that aim to optimize, regulate, and police all sorts of transgressors against proper human ways of living and dying (Stevenson 2014; Zeiderman 2013). Instead, they obliged me to depart from a biopolitical register that starkly distinguishes between life and death and the well-being of human life and the lives of other beings and things to make an ethico-analytical move that pays careful attention to processes of human and nonhuman material composition and decomposition. The agrarian-based material and ethical practices they shared open up ways of thinking about the different kinds of political, economic, and ecological relationalities that emerge when life and death are no longer or never could be experienced as oppositional categories or morally dictated ultimates. I am not suggesting that there are not undignified experiences of death, such as the violent dismemberment of victims' bodies by paramilitaries or state illicit crop eradication campaigns that rip out criminalized plants, leaving gaping holes in soils and rural families with no economic sustenance. However, neither does a dignified life depend upon productivist and chemically manipulated modes of optimization or the eradication of ways of life under the guise of moralized and universalized betterment.

The networks of seed guardians, community-organized agro-environmental schools, and alternative agricultural practitioners that I have introduced in this chapter seek to create conditions that resist rural displacement in a context of ongoing uncertainty. They do this by working toward a future that they may never see.[7] Padre Alcides engaged in recovering seeds and innovating agricultural and forestry practices in the present inhabited by a future territory that did not yet exist, but that continues to strive to humbly materialize its collective affordances and capacities. Similarly, regional social movements struggle to implement an Andean-Amazonian Integral Development Plan (PLADIA 2035) that aspires to structurally transform current antidrug policy despite the systematic breach of agreements on the part of the state. Dispersed networks of rural families engage in everyday practices to protect seeds, recover plants and recipes, and convert their gardens into successional agroforestry experiments without guarantees of lasting permanence. This being said, past territories were never fully eradicated because they remain present in different modes of cultivating and socializing and in the persistence of recipes, remedies, and criollo seeds. At the same time, the cultivation of a farm, forest, or garden strives to farm itself a community of future human and nonhuman cultivators.[8] I am reminded of Elizabeth Povinelli's discussion of potentiality as the "enduring fuel of a counterlife" (2011a, 128)—the extension of something over space and time that has no viable language outside the iteration of persistence or becoming of something different. Within the folds and recesses of potentiality are germinating actualizations—the creative emergences and actual work occurring in the present when cultivating different conditions for life and death constitutes a present and future contestatory act. Social movements in Putumayo describe this in terms of "propositional aggression" (MEROS 2015, 14), or being motivated by a rage that opposes and contests intolerable social conditions, and in doing so also works to cultivate alternatives to the asymmetric forces that make certain worlds thrive at the expense of forcing others to endure. This requires a delicate balance between opposing, *aguantar* (enduring or carrying on) within intolerable conditions, and actualizing alternatives to these very conditions.

Trece mil rayos tiene el sol.
Trece mil rayos tiene la luna.
Trece mil veces se han arrepentido los enemigos invisibles o visibles que
    tengo yo.
En el nombre del padre, del hijo, del gran espíritu
Gran padre, Gran chamán ancestral.
Así es.
Así es.
Así es.

Thirteen thousand rays has the sun.
Thirteen thousand rays has the moon.
Thirteen thousand times my invisible and visible enemies have felt
    regret.
In the name of the Father, the Son and the Great Spirit
Great father, Great ancestral chamán.
Thus it is
Thus it is
Thus it is.

After being violently dispossessed of five farms over the past twenty-five years, Heraldo now lives fifteen minutes down the road from Nelso and Elva. One morning while we were waiting for more passengers to board a public bus, Heraldo flipped through a tiny battered green notebook. He stopped on a page that revealed a prayer scrawled in pencil (see box). It was given to him the last time he visited his neighbor, the *taita* Don Lucho. "Have you memorized it?" I asked. "Not yet," Heraldo told me. "Does it help make you invisible?" I inquired, taking the word *invisible* in the prayer far too literally. "No," he said. "Rather imperceptible, undetectable, unfixed in time and space." In case this had caused me any confusion he added more concretely, "to avoid problems with thieves, paramilitaries, soldiers, nature, the body." That same morning before our trip, we received news that one of his neighbors had been gunned down at dawn. Upon hearing this, Heraldo explained to me, "I can die, but the Amazonian farms will remain." By this, he was not referring to the fixed desire to demarcate private property inhabited by a self-interested household, defor- ested fields, and ordered rows of commercial crops. These kinds of farms that fall in line with conventional state-based forms of property acquisition can easily slip away, and often work against the continuation of rural life in the ter- ritory. As I build upon in the next chapter, Heraldo was referring to cultivating conditions that "recolonize the farm with selva." This recolonization of farms by selva materially and conceptually disrupts notions of private property be- cause farms are never only farms when they are also always regional water- sheds, foothills, forests, ecological corridors, floodplains, and places of learn- ing, articulation, and exchange. I came to think of *farms as more than farms*, as the multiplication of pulsations of cosustainability: the different birds and bees attracted to creeping plant gardens, the reawakening of microbial worlds as communities of plants return, the multiplication of diverse selva practices resonating throughout a territory in community networks of seeds, the barter- ing of food and products, the building of campesino markets, the recovery of food autonomy, and the creation of conditions for the successional recycling of hojarasca—this layer of fallen and dying leaves that undergoes natural pro- cesses of decomposition, and when incorporated back into the germination of the earth is always already regenerative of selva life.

1,000 years parent rock appears,
100 days soils are made,
third reap soils deplete,
from selva to sugarcane, coca, and cattle,
soils retire and aspiring capitalists lose the battle.

As Heraldo's earlier prayer conveys, being rendered invisible is quite different from the subtle, processual nature of imperceptibility. The latter may allow for the avoidance of becoming fixed in any singular identity or perilous state. In an attempt to reflect carefully on the germinating withdrawal of "forgotten" seeds and the decompositional modes of retreat of deforested soils, I take inspiration from discussions of corporeal vulnerability. These discussions are occurring at the interfaces of feminist and critical disability studies and the environmental humanities on such phenomena as susceptibility, indolence, and fatigue (Butler, Gambetti, and Sabsay 2016; Diprose 2002; Harrison 2008; Jaquette Ray, Sibara, and Alaimo 2017). Rather than conceive of these states as degraded ways of being in the world that must be set right or transcended by benevolent action, fatigue can be understood as an occurrence whose reality is made up of refusal or hesitancy. One of the most poignant lessons Nelso, Elva, Heraldo, and others taught me is that the region's soils—and, as I hope has become clear, by soils they are never referring to a stable entity, but rather to continuous relations of composure and decomposure that in fleeting, massifying moments produces a natural body scientists call "soil"—could never be colonized, but only destroyed. In other words, Amazonian soils-selvas could not be tamed by human-led productive fantasies, but they could be used up and abandoned.[9] Similarly, all kinds of "subproductive" and uncertified seeds are not physically going extinct, but rather have become strangers, receding into the background to take refuge among inconspicuous "weeds" that covertly propagate alongside the agro-industrial economy. In a parallel manner, many community and agrarian leaders found themselves obliged to shift organizing strategies by retreating into anonymity in order to survive the war. It is through this seeming disappearance that a vocation to live might resist violent modes of death by becoming into it, which is not necessarily death or dying, but a melding back into an imperceptible and transformative respite, a sinking into injurious conditions, open wounds, and fatigue. Imperceptibility allows for engagement in struggles that are more akin to stamina, a propositional endurance composed through the flight of withdrawal, rather than activity writ large or ostensible confrontation that may or may not end in ultimate triumph.

According to Nelso, what many agronomists and conventional farmers describe as the "limiting fragility" of Amazonian soils is actually a mode of resistance that may allow—indeed oblige—rural communities to simultaneously slow down as well. It is not soils' intentional refusal in the same way that they express their own growing recalcitrance before the commodified notion of life underpinning growth-oriented economic imperatives. Instead, it is an expression of a capacity for weariness and recoil from the impossibility of existing under the relentless strain of extractive conditions that exceed the soil's abilities to absorb, repose, and transform. Similar to Butler, Gambetti, and Sabsay's (2016) discussion of resistance as drawing from and mobilizing vulnerability as a necessary condition for action rather than simply being a debilitating form of exposure and precarity, these families taught me that what appears dead is not. For Nelso and Elva, "playing dead" marks the soil's imperceptible and regenerative unworking and reworking that evades not only its own exploitation, but also a continuum of exploitation linking rural families, microbes, plants, seeds, soils, and trees—not simply a mechanical cause-and-effect interaction or biological breakdown, but the sheer physical force of fragility, the feeding loop of activity and withdrawal characterizing an ecological relationality. Relatedly, Heraldo and Nelso argue that when technicians and bureaucrats claim that one cannot sustain a productive agricultural livelihood in the Amazon, what they are really saying is that one cannot sustain a colonizing, extractive, and neoliberal agricultural livelihood.

In this book, I am not telling a triumphant story of small-scale agroecological farming as an improved agricultural model that should be singularly adopted throughout Colombia's western Amazon. Nor do I want to argue that technical decisions over crop choice and livelihood are always or simply a political choice.[10] My intention is to highlight the built-in expectation of vulnerability present in an ecological sense of being in the world: a family's necessary acceptance of decay, looping back to begin again, a cautionary lying low, and a decompositional mode of regeneration. These alternative agricultural practitioners remind us that transformative potentiality is not a human privilege, but always a relational matter dispersed in the connections and labor among people as well as other beings and things. They also urge us to take seriously the ways particular human and nonhuman relations afford differential political and economic capacities. Similar to Povinelli's (2011b) discussion of self-transformation, any notion of collective determination cannot be separated from a host of relations with/in place. This includes material transfers—

eating, pissing, shitting, and sweating—that send matter back into the recycling interplay that composes soil-selva life and death.

At stake in these rural communities' struggles is not the right to idleness, but the right to another kind of work, another kind of dream, and a world that does not run only on the inevitable and structurally designed market-based time and velocity.[11] What may be most inspiring is how their agro-life processes slow down the power of dominant agrarian reasoning to create a space for alternative practices and creative experimentation: for example, planting "weeds," living in the midst of animal fodder, preferring not to "correct" or resuscitate "poor" soils, rejecting the salvation narratives of modernizing agricultural sciences or state antidrug campaigns that pursue peace through poison. Much as Stengers (2005a) does, their practices demand a slowing down of the power of reasoning imbued with dominant concepts. This slowing down propels them to ask what claims to knowledge are being made when certain practices are declared to be backward, obsolete, and/or criminal. It is the speeds of nutrient and metabolic cycles and erosion patterns, the intensity of rainfall, and the proliferation of microbial life in the Amazon that slow down the velocity through which "problems" (i.e., poor, acidic soils) are formulated as such, and then promptly resolved while rendering unimaginable other kinds of thought and action. The acceleration of ecological processes in the Amazon inspires a kind of deceleration, which slows down the "reason" attached to dominant narratives promoting industrial agricultural technical packages, capitalist market values, ethical approaches, and temporal inevitabilities.

Unlike the agrologists I presented in the previous chapter, this deceleration does not imply the need to solve troubling "enigmas." A growing number of rural communities argue that there is something more important at stake than uprooting illicit crops or biopolitically improving the productive capacities of soils and souls—something more vital than producing enough exportable commodities to feed a far-off world whose moral standing is protected by continued assaults against the worlds that these communities inhabit and strive to dream and build. They suggest that the wrong kinds of questions are being asked. Perhaps it is not a question of becoming better, more productive farmers, more legal and more responsible citizens, or more efficient capitalists. It is necessary to slow down and ask what kinds of questions emerge from an ecologically relational world that not only obliges different strategies for how to keep on enduring in the face of a war machine that proposes peace through poison, but that strives to cultivate different socioecological, economic, and political realities.

These struggles in southern Colombia articulate with larger contemporary political processes in the Global North and South, including anti-extractivism, anticapitalism, transitional initiatives, and degrowth (D'Alisa, DeMaria, and Kallis 2015; Klein 2014; Roy and Martínez Alier 2017). Throughout the Andes, they also resonate with heterogeneous practices of the Quechua-Aymara concepts of *sumak kawsay* and *suma qamaña*, or what has often been translated into Spanish as *buen vivir* (living well), which reject human-centered notions of growth and productivity tied to a universal *vivir mejor* (living better) (Gudynas and Acosta 2011). Living well is, of course, never separate from the question of creating the conditions for dying well—of the right to form part of processes through which death decomposes into life rather than being violently ripped from place, territory, soil-selva, and home. This is a mode of death that decomposes into life much as leaves fall from branch to ground, turn over, and rot to germinate from a pulsating layer of hojarasca again.

BORDER CROSSING

Think on your feet, the woman sitting next to me instructs.
Her voice is a rough whisper scratching against the air.
Before the paramilitaries stop the boat, hide the dollars in your underwear.
All greenbacks are assumed to be destined for the guerrilla.

The last time she crossed the river from Ecuador this saved her life.
$200
A thick wad of bills hastily pressed against a warm groin.

4 men shot.
4 men dead.
And one of them was only carrying a $20.

She states the facts, simply.
Know the rules even when they change.
Putting on underwear may be the most important thing you do every day.

5

RESONATING FARMS AND VITAL SPACES

A Person and His Concepts

INSIDE THE CHRYSALIS

Heraldo would sometimes find me with my head cocked to one side, ears perked in anticipation, standing motionless in the middle of one of his Amazonian gardens. This was my favorite spot, home to the *enredaderas* or *bejucos* (creeping plants), the place I always sought out whenever I arrived and did not initially find him because he was at work in some more distant corner of the farm. Each time I walked down the footpath from the main road and felt myself engulfed by the light perfume of the wild *orquidea* flowers that Heraldo planted at the entrance as a kind of welcome mat, I never knew what changes I would encounter upon rounding the corner to find the first visible view of the farmhouse. There were small buildings under construction, piles of materials, holes dug in the ground, and seedling trees that would eventually provide shade for the additional structures. For most of my fieldwork, Heraldo was in the midst of designing and constructing an open-air classroom, two composting outhouses, a dormitory, and a large brick oven to prepare food for future visitors, students, volunteers, and customers. He also constructed six polyculture fishponds; a dehydrator to dry seeds, grains, and beans to make different kinds of flour and animal feed; a pen for the guinea pigs whose manure he used to fertilize the farm; and a sugarcane press to produce raw cane sugar for human and animal consumption. Heraldo's goal is to no longer solicit or accept government and nonprofit contracted jobs once he finishes building the farm school. Since its inception, the farm has been free and open to anyone interested in learning alternative Amazonian agricultural practices. Over the years, he has hosted a series of technical workshops, seed exchanges, community meetings, and political gatherings. His plans are to barter, sell, and transform a portion of the harvests and to prepare Amazonian recipes and products for townspeople who are interested in purchasing lunch and roasted guinea pigs for special occasions, a delicacy brought by people who migrated to Putumayo from the neighboring state of Nariño.

The farm, he often told me, would never have a sign or official name because this would require that it be registered and subsequently taxed by the municipal chamber of commerce. If it were registered, officials from the National Institute of Medical and Food Monitoring (INVIMA) would likely arrive to regulate his food preparation, packaging, and commercial labeling practices. This would lead to visits by officials from other government regulatory agencies that are designed to standardize and control industrial-level agricultural production without offering differential policies to protect and support the particularities of campesino and small farming livelihoods and economies. It is precisely Heraldo's outside income that provides him with the resources to build a small-scale integral Amazonian farm school. Of course, he is aware that contractual employment, albeit irregular, is unavailable to most rural families who have not had access to formal education and similar technical training. As I have explained, integral Amazonian farm schools are not intended to become a replicable technical model. They are instead a place of learning and exchange, a space of articulation. The experimental practices, seeds, strategies, and recipes learned and shared are necessarily modified from one farm to the next following the situated agroecological conditions, aspirations, and material resources of each rural family, organization, network, and, as I will expand upon, associative economic structures. Heraldo himself works three kinds of "sustainable agricultural arrangements": (1) associated farms or mixed forests, such as agroforestry, silvipastoral, agrosilvipastoral, and silvicultural arrangements; (2) integrated or integral farms that recycle relations between two or more activities, such as raising guinea pigs and worms and integrating their manure and compost as nutrients for the grasses, trees, and plants; and (3) what he calls multiple farms, where several systems are intermixed, such as his agroforestry system associated with other integral practices.

Heraldo bought these two hectares of land in the vereda Rumiyaco four years before we first met. At the time, they were open pasture full of the kinds

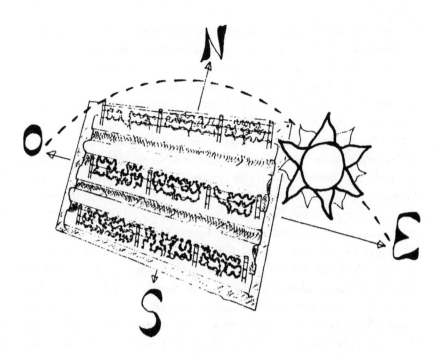

FIGURE 5.1 Diagram of how to build a garden following the sunlight, so that no plant overshadows another (Clínica Ambiental 2009, 1). Photograph by author.

of invasive grasses that thrive in degraded soils after thirty years of monocrop sugarcane production followed by the compacting hooves of one too many cattle. It was his fifth farm, his fifth "attempt," as he put it. The other four farms in Valle del Guaméz, Villagarzón, and the Media Bota Caucana had been lost to a mix of bad partnerships, thievery, and military-paramilitary occupations at various moments over the last twenty-five years. Heraldo did not dwell on the details of how each of the farms had been lost or violently expropriated, and standing here with him in the middle of the enredaderas felt like a tenuous achievement. There is a sensuous intensity to the way the creeping plants wind themselves around the rusty barbed wire and wooden posts, sprawling out like green braids speckled with fruits, vegetables, and nuts. Working in the Amazonian gardens and orchards with Heraldo entails being quick on one's feet. He moves swiftly, weaving in and out of the rows and attending to the gnarly mass of plants and beds of tubers growing below—garlic del monte (an Amazonian alternative to conventional onion), spinach del monte (an alternative to Andean lettuce), star nut, wild passion fruit, air potato, *torta* beans, *ba-*

dea, huasca tea leaves—inspecting for worms and fungus and removing them manually. He looks for unfavorable signs that the plants are, as he says, "eating off each other's plates and hence all going a little hungry."

Heraldo took these opportunities to remind me that every organism needs what he calls an *"espacio vital"* (vital space), borrowing from the Brazilian-based agroecologist Ana Primavesi (1984), in order to keep on eating together—that is, from and with each other, but not against each other. In the most technical sense, he was teaching me about different kinds of garden eating symbioses. To avoid excessive parasitism and competition between plants one should sow plants of the same species farther away from each other and intermixed with diverse companions that have deeper-penetrating and shallower roots. This enables a redistribution of a wider range of macro- and micronutrients available in the different horizons of the soil. Plant and crop selections also depend on the hours of direct sunlight in a given locale. Heraldo mapped this out for the different subregions of Putumayo, starting from two hours of direct sunlight a day in the Andean-Amazonian foothills to a gradual increase to four hours a day when reaching the Amazonian plains of the municipality of Puerto Leguizamo at one of the southernmost extremes of the state. The hours of available direct sunlight contribute to his argument against the implementation of plantation-style African palm oil in the Amazon because these trees require at least five hours of direct sunlight a day to reach their optimal production according to the industry's own standards.

This particular afternoon, Heraldo chuckled upon discovering me in the garden, I think because he realized that I too had been drawn in, enredada (entangled) in the sensorial density and textured materialities of the creeping plants. Standing in the midst of the garden was like swallowing the first gulps of air from inside a chrysalis. I felt my body gradually expand, the silky casing around me about to give way. When he asked me how I felt, I responded almost instinctively. *Like a butterfly*, I said, *an emergent butterfly slowly inflating her wings.* Heraldo was pleased because my response encouraged his idea to amplify what I began to call the *vital frequency* of the garden by building *parlantes naturales* (natural speakers) that would reverberate across and beyond the farm. He contemplated whether to hang plastic tubes in the orchard to amplify the frequency of the buzzing insects and bird calls, and he told me that the late afternoon breeze would play an important circulatory role. If only he could magnify the sounds, the vibrations, the life of the gardens, he said, then imagine how the neighbors might react; how the cattle living without shade would feel; how la selva might respond.

The way I imagine it, over time, dispersed farms across the Andean-

FIGURE 5.2 *Papa aérea*, or what is otherwise known as *ñame volador* (air potato), growing in the creeping plant garden, March 2011. Photograph by author.

Amazonian foothills and plains might begin to vibrate, buzz, and hum. This could produce a cacophony of vital frequencies—life making life happier—as opposed to the quieting exhaustion and laments of campesinos overwhelmed by debt, bad-faith politicians, exploitative intermediaries, and the market prices and mandates of armed actors, narco-mafia networks, or externally imposed aid programs. It was not a question of how many farms might come to resonate because not every rural family could or should be moved—that is, become capable of being affected and transformed—by the resonating throb. Returning to my description in chapter 1 of the "noise" generated by La Hojarasca farm school in the midst of a militarized development-intervened landscape, what is at stake in this vital frequency is the espacios vitales that might be created and defended. Borrowing from Ana Primavesi's technical discussion of the necessary space between plant species in any agroecological arrangement, I began to think with espacios vitales as a conceptual and political tool from which to imagine what an Andean-Amazonian territory of resonating farms might look and sound like. Scaling up, I became interested in the question of how a territory might become a vital space. In this chapter, I focus on the conceptual work that accompanies a "territorial turn"—or better yet, territorial opening—among rural communities across the Andean-Amazonian foothills and plains—in particular, the kinds of conceptual personages that form part of selva agro-life processes and the role of garden dreaming in imagining and plying a different material reality.

TALKING FARMS

When I listened to Heraldo explain his selva agricultural proposal to rural communities during popular workshops and community meetings, he often said, "The Amazonian farms will talk." More importantly, "they will talk on their own," meaning that they would not need a human spokesperson to convince anyone to adopt this or that alternative technical model. What I came to think of as the nonrepresentational aspiration of resonating farms was intended to free up rural communities from the regional *agropolitiquería* (agropoliticking). Farms would talk so that these communities would no longer be obliged to engage with the local and regional electoral political machine, an activity that pulls them away from life and labor on their farms and renders them dependent on the false promises, corruption, and clientelism of individual candidates and traditional political parties. What I call resonating farms do not intend to become an independent social or political movement, but rather spaces of articulation with the ability to network pluralistic alternative agricultural initiatives on the part of campesino, indigenous, and Afrodescendant communities. According to Heraldo, the emergence of a new Amazonian "agroecological" social movement, for example, could likely fragment rather than strengthen already existing and potential partial alliances across ethnically diverse rural communities whose social fabrics have been fraught by years of war, assistencialism, and official multicultural policies.

State multicultural legal frameworks have often served to pit rural communities against each other along ethnic and cultural lines while denying campesinos legal protection as an economically or culturally differentiated group. Carlos Duarte (2016) and others at the Javeriana University's Center for Intercultural Studies argue that *desencuentros territoriales (*territorial disagreements*)* are, in part, a consequence of the way multicultural constitutional reforms in 1991 introduced a differential and asymmetric framework of rights and social protections that guarantee the cultural integrity of indigenous and Afro-descendent populations, even when the majority of these groups live in rural worlds shared with campesino and mestizo communities enduring similar or overlapping experiences of social exclusion and agrarian-based violence. Multicultural legislation in Colombia recognizes three types of collective territory or collective land titles: indigenous reservations (*resguardos*); Afro-descendant community councils (*consejos comunitarios*) (Law 70 in 1993); and campesino reserves (*zonas de reserva campesina*—ZRC) (Law 160 in 1994). Currently, there are six officially constituted ZRCs in Colombia despite the existence of many more community-led processes to constitute reserves. ZRC processes have been systematically impeded by the state due to their presence in territories that were historically occupied by guerrilla organizations or due to interethnic territorial conflicts between rural communities.[1] Multicultural policies have been unable to attend to existing complex and heterogeneous interethnic territorial configurations as well as newly emerging, ambiguous, and transitory territorial arrangements, such as legally unrecognized yet community-declared campesino reserves; agrofood territories (*territorios campesinos agroalimentarias*); and rural communities that are politically constituted but that have no collective territory. These conflicts demonstrate the necessity for alternative modes of intercultural dialogue and agrarian jurisdiction. It is crucial to emphasize that while differential access to constitutional rights and the collective autonomy of ethnically identified communities is an important focal point of analysis, the principal vectors of territorial conflict are derived from structural conditions: poverty, social exclusion, and a deeply rooted land tenure model that has sought to concentrate land in the hands of latifundium cattle-breeding and agro-industrial sectors at the expense of minifundium rural communities (Duarte 2017).[2]

The transformation of what I have called *farms as more than farms* among rural communities can be situated within a context that Maristella Svampa (2012) defines as the "consensus of commodities" across Latin America. This consensus is based on the scaling up of projects aimed at the large-scale exploitation and export of natural resources, provoking what she calls an "eco-territorial turn," which has become the hallmark of social and campesino movements objecting to such projects. Agrarian-based movements throughout the Americas have learned important conceptual and political lessons from the territorial thinking and ethical commitments of the long-standing struggles of

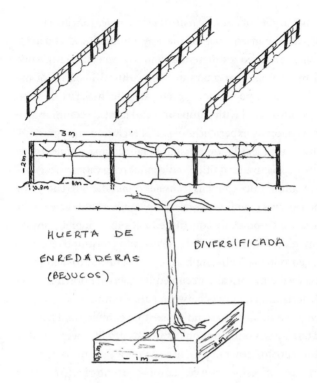

HUERTA DE ENREDADERAS (BEJUCOS)

DIVERSIFICADA

FIGURE 5.3 Heraldo's drawing of a creeping plant garden. Photograph by author.

their indigenous and Afro-descendant counterparts. Popular exercises of territorial ordinances conducted by campesino organizations across the hemisphere are increasingly conceptually and materially informed by a "territorial shift" in which historical demands for land and the right to property are accompanied by a broader defense of life and territory. This has led a growing number of communities to conceive of and organize their farms not only as economically productive spaces, but also as connective places with ethical obligations to myriad socioecological continuums that both incorporate and extend far beyond the property or occupational limits of the farm: farms designed and defended as watershed and wetland areas, integral gardens and orchards, transition zones between foothills and plains, and wildlife and biological corridors.

Interestingly, Heraldo's proposal of "talking farms" also takes conceptual cues from the affective workings of coca plants, specifically, the emotive ties that these plants have been capable of garnering among rural communities— envy, hope, boldness, risk, and desire—and the way coca bushes sprouted up and transformed entire regional economies and socio-ecological relations, multiplying until even the most reticent individuals found themselves cohabiting if not co-laboring with the plant.[3] He often commented on the energies harnessed by coca, less in its global commodity dimension and violent associations than in its ability to set new relations and affects in motion. We had many conversations about the mobilizing potential of *afecto* (affect, attachment)—not in the conventional political sense of organizing people into consolidated grassroots movements or political parties, but in the ability to get hearts stirring, dispersed initiatives on the move, and, as Heraldo said, folks " *que están pensando entre más que una cabeza*" (that are thinking between more than one head).

Heraldo's reflections on affect are informed by his participation in leftist student movements at the University of Nariño in the late 1970s and early '80s. Many of his classmates and friends later went on to join different workers' and communist parties while others became members of the M-19 (19th of April) guerrilla movement. Heraldo does not agree with the use of armed violence or what he calls "sectarian politics" as a means to build alternative life processes; however, he was inspired by what he perceived to be M-19's focus on pressure points, relationality, and love, and not just orthodox Marxist political ideology. He takes seriously how the stirrings of desire provoked by coca plants, which have been capable of inciting new, albeit monoculture and extractive-based human–plant relations, economic subjectivities, and ethical horizons, might be inherited and channeled toward altogether different agro-life processes. Indeed, how might life, as Heraldo posed again that afternoon amid the enredaderas, make life happier—not better, but happier? How might happier farms expand rather than replicate one another, each having the potential to become something of its own with the capacity, but never the obligation, to inspire the workings of another?

What I am calling resonating farms is not a realistic project, but rather a processual unfolding of dispersed efforts to compose and decompose into a different Amazonian territory. If hope, borrowing from Stengers, "is the difference between possibility and probability" (2002, 245), then we might conceive of resonating farms as practicing possibility in a world disciplined by discourses of probability. Probability materializes through acts of brute violence and dispossession, in the alienating and hierarchical technical languages of expertise, and the supposed inevitability of capitalist market imperatives and temporalities. Inspired by Whitehead, Stengers speaks of hope as an attempt "to feel what lurks in the interstices" (245), or that which usually escapes description because the words most readily available to us most often refer to stabilized identities.[4] Practicing hope is, of course, a risky undertaking in the face of the banal and conspicuous forces that work to homogenize, eradicate, and deny other realities and regional imaginings of belonging, inhabiting, and becoming. As I have described in previous chapters, it is not only rural families

FIGURE 5.4 Caterpillar larvae munching on a granadilla leaf in Heraldo's creeping plant garden. Mocoa, Putumayo, March 2011. Photograph by author.

in coca-growing regions of the country that are attempting to escape stabilizing and stigmatizing state categories, but also regional soils, which are trapped in a hierarchical classificatory system that deems them impoverished, prone to criminal activity, and in need of "correcting."

The afternoon that Heraldo drew this table (table 5.1) he was conversing with a group of campesinos who had traveled to his farm school from Bajo Putumayo. "One plant plus one plant does not equal two," he said, referring to the linear logic of mass production. "*Más* [more] is said to signify *más plata* [more money] and *más producción* [more production of plants of a single species over a designated surface unit] instead of *más diversidad* [more diversity] actually contributing to better production." In contrast to the Ministry of Agriculture's concept of a chain of production, which Heraldo conceives as a linear *hilo* (thread), he proposes the idea of *mallas* (meshes or nets) to conceptualize what I think of as regional agro-food-cultivating-eating-shitting cycles. Meshes convey the complexity of agrobiodiversity as more than a group of diverse living beings that share a designated and obligated space. As Heraldo explains it, complexity results from systemic diversity at different trophic levels that interact to function as sustainable ecological rotations with their

TABLE 5.1 Heraldo's Chart for Defying Linear Logics

Monoculture: deifying production for exportation	Polyculture: vital spaces, biotic communities, fewer plants, but each "produces" more foliar area
Improved seeds	Native or adapted seeds
Measuring the soil or defining it in terms of fertility	Understanding that the "soil" is a product of the agricultural arreglo (relational arrangements)
NPK in chemical form	NPK produced by the dynamic associations between plants, atmosphere, and microbial ecology
One form of understanding "wealth"	Another form of understanding "abundance"

own inputs and outputs and as ecosystems unto themselves (Vallejo 1993b). For example, many rural communities in Putumayo criticized USAID agroforestry models, which tended to group rubber and plantain trees with black pepper plants under the auspices of biodiverse agroforestry, because there is no systemic level of ecological interaction in this model, such as accompanying plants that produce inputs for inhabitants of the farm other than humans, as well as outputs for the human family. Heraldo's mention of "better production" does not refer to the more efficient and intensive accumulation of economic wealth, but, as his table indicates, more ample notions of abundance that include diversified and solidarity-based economic relations as well as the proliferation of more foliar area and vital spaces. In other words, abundance is understood as the proliferation of difference.

I am led to place Heraldo's "linear defying logics" and the agro-life processes they are attached to in dialogue with the "anti-production machine" that Pierre Clastres (2007, 2010) reminds us permanently haunts the productive forces. Clastres refers to the underproductivity or insufficiency that pulsates beneath the moralism of the economy and that inhibits accumulation precisely by expressing a sociopolitical necessity of nonexchange. Lévi-Strauss calls these orientations "savage thought," not because they are the thought of savages understood through the racialized evolutionism that leads from savagery to civilization or from states of nature to civilized states, but in the capacious potential of all thought as long as it is not "domesticated for the purpose of yielding a return" (1966, 219). Throughout the Andean-Amazonian foothills and plains, I accompanied families striving to reestablish practices of bartering, borrowing, gifting, and cooperative work (mingas). They struggle to resist the fixing of equivalency, commodification, competition, and accumulation as the founding principles of agri-sociality. People's attempts to enact what J. K. Gibson-Graham conceived of as "landscapes of economic difference" (2006a, ix), or breaks in the performance of dominant national and regional economic relations, are evidenced in community-based seed certification networks, the elimination of packaging and commercial labels, artisanal food production, recycling and bartering, and the refusal to register and pay taxes on commercial activities. These are examples of what Gibson-Graham called "fugitive energies" (2006b, 51) and affects that exceed institutionally provided and assumed subjectivities and enactments of labor and transaction.[5]

Of course, political economies of escape are not purist exercises or totalizing projects. Autonomy is always transitional, relational, material, and aspirational.[6] The alternative agricultural practitioners I accompanied deeply tie economic autonomy, especially food autonomy and sovereignty, to political autonomy. They describe food security as a government's obligation to guarantee access to quality food to all its people. What quality comes to signify is, of course, a disputed issue. Food sovereignty is conceived as an integral component of the protection of national sovereignty when a government has the capacity to guarantee autonomous and sustainable food production within its borders and to safeguard the country's resources from domination by foreign powers. More than market competitiveness, this second point speaks to the sustainability of national food production. Campesinos described food autonomy as the free determination of all communities, peoples, and rural families to produce, prepare, and consume the food that is compatible with their preferences and territorial agroecological conditions. This implies that they have autonomy—collective determination—over their territories and agrarian policies (Lyons 2016b).[7]

STARVED SCIENCE AND ALTERNATIVE AGROECOLOGIES AND ECONOMIES

In 2010, I attended a roundtable organized by the municipal secretaries of health and social issues for the purpose of formulating Mocoa's municipal food security plan. This is a plan that each elected mayor is obligated to submit to the governor, who is then required to submit a statewide plan to the federal government. The few campesinos who had been invited to participate in this particular event argued that food security "fills one's stomach but leaves one's soul hungry." They viewed the food security workshops and policies organized by state nutritionists as not only ineffectual, but even worse; as one of the leaders of the Federation of Afro-Descendent Communities of Putumayo

(FEDECAP) explained, "these policies malnourish the people." The idea of food security promoted through institutional mechanisms is measured in terms of the availability of and access to "quality food and effective nutritional standards" that meet the designated *canasta familiar* (family food basket). However, the charts that nutritionists show communities outlining the percentages of vitamins and generic food groups they should consume has little to do with the specific plants and vegetation growing in the Amazon. "Hunger is integral," Nelso said at the roundtable meeting. "With the state's concept of food security, one is not necessarily eating *here* after walking the selva and visiting people's gardens, learning from plants and seeds, and making the recipes they allow and inspire."

Over a span of several years, Heraldo, Nelso, Elva, and other alternative agricultural practitioners and local professionals and technicians in Mocoa formed a community network called SOCIVIL (Civil Society of Putumayo). Between 2008 and 2010, SOCIVIL organized four regional forums with different rural communities called La Cuchara [The Spoon]: For Food Autonomy, Sovereignty, and Security in the Andean-Amazonian Corridor. The goal of these forums was to create a space for collective discussion and analysis of the precarious situation of food autonomy in the region—in particular, the overwhelming dependency on food imports from neighboring regions, the consumption of products containing high levels of agro-toxins, and the environmental impacts of agroecologically inappropriate agricultural and livestock practices. The image chosen to symbolize La Cuchara was a large traditional wooden spoon. "State officials say that we want people to eat monte [wild plants, animal fodder], while they expect people to be nurtured by a starved science," said Nelso at one of La Cuchara gatherings. The reality that rural communities have become consumers rather than producers of food renders them vulnerable to what Nelso and Elva call "*La Almuerzocracia*" (which I translate in an equivalent play on words as lunchocracy). They refer to the way political candidates offer free lunches, snacks, and other tokens to capture votes during election season. "*No bote tu voto*" (Don't waste your vote). "Not for a roof tile, a contract, money, a soft drink, or even a lunch!" Don Nelso warned his listeners on his community radio show when national and local elections neared. Similarly, the fact that coca growers have practically stopped growing subsistence crops has left them completely dependent on the market prices paid by narcotraffickers and their intermediaries. This was one reason why certain fronts of the FARC-EP obliged the families living in the territories they occupied to grow a minimal amount of subsistence crops.

While many of the campesinos I met argue that they are tired of having "a

FIGURE 5.5 Toasting sachainche (star nut) in a SOCIVIL forum with La Cuchara. Mocoa, Putumayo. Photograph by author.

dollar sign branded to their foreheads," this is not because they think they live in a world where they no longer need money or economic solvency or because they reject all market transactions and state social investment. What the state denominates as illicit economic activities is both the underbelly and the margins of the global capitalist economy. While many social scientists and human rights reports place emphasis on the state's historic abandonment of frontier zones such as Putumayo, it is not that the state has been absent, but instead that its presence has been limited to militarization, antinarcotics operations, and the intensification of extractive industry. USAID and other development and aid operators, multinational oil consortiums, FARC-EP guerrillas, and local communities have built much of the existing infrastructure in the department. However, while social movements denounce what they perceive to be the state's historic socioeconomic and environmental debts to the region, they are not interested in just any state institutional presence or education, health, housing, agriculture, and development programs, much less those that are conceived without the full participation of local communities and the agroecological specificities of Amazonian realities.

The Andean-Amazonian Integral Development Plan (PLADIA 2035) crafted by MEROS proposes a regional economic reconversion or transitional approach that envisions markets in terms of layered scales. PLADIA proposes that rural communities first recover their local and regional food autonomy, agrobiodiversity, and multipurpose and economically pluralistic campesino markets. The design of each integral agro-productive farm includes establishing agricultural arrangements of commercial crops by building associative networks between families, since no one family will engage in monoculture intensive production of any single crop. The proposal is to couple economic associativity with innovation in transformation. In this way, farmers will no longer sell raw materials to an intermediary, but will instead transform their crops, such as making sachainche oil, Amazonian chocolate, coca flour (*harina de coca*) and other coca-based products, and cocona and a range of Amazonian fruit juices and pulps, to sell directly to buyers. Market scaling begins by reconnecting producers and consumers throughout the veredas and towns of Putumayo's municipalities, and then building more equitable and solidarity-based exchanges between Andean and Amazonian producers, consumers, and markets. The eventual surplus production of local and regional products would then be integrated into domestic and international niche markets.

Other communities in Putumayo have started cultivating agroforestry projects or *bosques comestibles* (edible forests). They propose receiving a subsidy over a ten-year period, during which time they would recover deforested watershed areas and also learn how to grow native timber-yielding and naturally regenerating varieties of Amazonian trees in sustainable agro-silvicultural systems. Still other groups of families have been experimenting with the cultivation of organic rice, utilizing no-till and no-burn crop residue practices in *chuquias* or terraced areas where water naturally dams up. Inspired by public policies in Bolivia and Peru, coca growers and workers have repeatedly proposed that the state buy a certain amount of legalized coca leaves and coca-derived medicinal, cosmetic, artisanal, and nutritional products. This latter proposal, which was presented during the National Agrarian, Ethnic, and Popular Strike in 2013, and at the National Constituent Assembly of Coca, Poppy, and Marijuana Growers in 2015, has thus far received no state support. I am drawn to Viveiros de Castro's discussion of the tensions between necessity and sufficiency to reflect on the aspirations of rural communities to transition out of extractive-based, capitalist economic practices, and the kinds of homogenization and dependency that this system has produced and perpetuates. Viveiros de Castro (2013) writes about the need to counter economic development—sustainable and otherwise—by generating concepts of anthropological sufficiency, or what he refers to as political self-determination: the capacity of a people to define for themselves what is a *good enough life*.[8] Afro-descendant and indigenous communities in Colombia call this *pensamiento propio* (a group's or collective's own thoughts) (see Escobar 2008; Rosero 2002). As I go on to explain, I heard campesinos in Putumayo conceptualize their thinking as *"un sancocho de ideas propias de aqui"* (a stew of ideas from here).

BECOMING SELVACINOS/AS

The only formal interview I ever attempted to conduct with Heraldo was during the last week of my long-term fieldwork. Like most farmers, he was a moving target. I had written down a list of my lingering questions, and I walked alongside, around, and bending down with him and two neighbors as they worked between several projects: the construction of a brick oven from scrap metal and a new pen with better shade for the guinea pigs. Heraldo was never particularly concerned with the precise dates of different occurrences in his life, and I quickly gave up trying to pinpoint this kind of information, much in the way that I stopped seeking to resolve the ambiguity surrounding the way people often narrated incidents of violence in the region. The most opportune moments to learn about Heraldo's life, and through his stories about life conditions in Putumayo since the 1930s when his grandfather migrated from San Lorenzo, Nariño, were during the first hours of our frequent road trips. After this window of time, he generally nodded off, and I was left listening to the salsa and bachata beats on the radio of the public transportation.

Staring out at the landscape running parallel to Putumayo's roadways one sees an unchanging view of open pasture that transitions into Andean foothills somewhere in the distance. Eggshell-colored oil pipelines at times run along both sides of the highway, literally blocking the way to people's front doors. The families living in these areas are forced to climb over the pipes to reach the road, passing toddlers, packages, and backpacks to open arms on the other side. Year round, women and children line the roadside with wooden stands where they sell plastic bags full of sliced white sugar pineapple and fruit of palm (*chontaduro*). I remember one repetitive scene in particular. A woman stands with a shovel in the middle of the unpaved highway, tiny piles of dirt and stones at her feet. She fills in the holes left behind by the weight of crude-oil trucks. Almost five hundred tankers press down on this road every day. When bus drivers, motorcyclists, or truckers approach on the road, she raises a thin string intended to pause traffic, just for a moment, one brief expectant moment. This woman becomes a reflection in the rearview mirror—dusty outstretched hand, sunburned face, and thinning piece of string. Rarely does anyone stop to pay for her public works.

Each time we headed south and passed the vereda Aguas Negras outside of Putumayo's commercial capital, Puerto Asís, where Heraldo spent most of his childhood, he pointed out his childhood home. His mother and youngest sister still live there, and he would tell me storied snapshots of his family's history. Like so many migrants to Putumayo, Heraldo describes his parents as hard workers, but conventional campesinos. Once their soils seemed to tire and yields of rice and corn declined, they transitioned to rotating parcels of acid-tolerant pineapple. When we passed a neighboring dilapidated farmhouse, I learned that the family living there had once been the largest pineapple business in the area until the group of sons decided to buy land farther south in the municipality of San Miguel to try their hand at coca. They were eventually murdered, and people now refer to their wives as the "pineapple widows."

It is Heraldo's paternal grandparents and great-uncle who take center stage in his early memories. His grandmother, Teresa, taught him to work in the garden and took him along to learn how to barter in lieu of buying food and supplies. She made *arepas de pan de norte* (an Amazonian alternative to corn arepas) that Heraldo points out have all but disappeared from the local diet. He describes his great-uncle Jerminas as somewhat of a recluse who lived alone in the monte reading all the newspapers he could acquire, especially the leftist *Voice of the Proletariat*. When he was younger, Jerminas traveled all over Colombia, and in his old age he took to writing the president and some of the ministers about Putumayo's socioeconomic problems. He dictated the letters to Heraldo, and a few times they even received a reply. His uncle never cleared the land, but planted "here and there," Heraldo said, beneath the forest canopy like many of his indigenous neighbors. Jerminas raised creole hens that eat Amazonian wheat and insects rather than commercial feed, and Amazonian fish that feed from local vegetation, such as *bore* leaves and stalks. "I realize he was doing what some people now call the science of agroecology," Heraldo told me.

Heraldo's grandfather, a staunch conservative, marched in all the civic protests of the 1960s to demand roads, ambulances, school desks, and other services and infrastructure after the territory was annexed from Nariño to become the Comisaría Especial del Putumayo. Comisarías were an older territorial subdivision that, along with *intendencies*, conformed to what were generically known as the country's "national territories" (*territorios nacionales*) administered under tutelage by a special office in the central government. Margarita Serje (2011) writes that these "savage territories"—low in population, generally indigenous, and very far from the capital and other urban centers of the country—were first converted into missionary outposts, then agricultural frontiers and fronts of colonization chronically problematic for the state, and then later became known as "zones of public order," the epicenters of the most intense violence of the country's social and armed conflict, and reduced to the status of pure representation by political elites. Sometimes Heraldo's grandfather escaped police retaliation during the protests. Other times he ended up in jail.

Like the majority of student leaders at the time, Heraldo spent six months in jail as a political prisoner in the city of Pasto due to his participation in the Revolutionary Youth of Colombia movement, one of the Marxist-based student-worker-campesino groups proliferating on public university campuses in the 1970s and '80s.[9] Amused, he told me that he was able to finish his undergraduate degree in animal husbandry while imprisoned because the prison pigs became gravely ill. Without syringes to give them injections of iron, the guards did not know how to remedy the pigs' severe anemia. Heraldo advised them to have the pigs eat and roll around in the patches of dark, mineral-rich soil in the prison yard. He wrote his senior thesis on the topic. Once released from prison, Heraldo returned to Putumayo after the state intensified its persecution of the Worker's Party (PTC) and other leftist groups, torturing, killing, and imprisoning members on the grounds that they were terrorists. Many of his colleagues were forced to disperse, join armed guerrilla groups, flee the country, or, like Heraldo, discretely slip back into the interstices of civilian life. There is nothing romantic or heroic about imprisonment. Going to jail, even for six months, which is not an enormous amount of time compared to the war already going on in the countryside for decades, expresses the threat of the state as it attempts to control those who oppose it. During my fieldwork, many social leaders and close friends in Putumayo were imprisoned for years on the basis of *montajes jurídicos* (false legal cases) that charged them with being terrorists or collaborators of the guerrillas. These montajes, as well as targeted assassinations, continue to be a strategy of the state and paramilitary actors to criminalize, smother, derail, and inspire fear in social movements and popular mobilizations.

There is no singular story to tell about how certain rural communities and families come to engage in what I refer to as selva agro-life processes. They read agroecology pamphlets and remember ways of doing things that their grandparents and indigenous neighbors performed. They exchange experiences, stories, and seeds with other rural families. They take long walks in the selva to study plants, trees, and animal behavior. Some have read Marx through their contact with the FARC-EP or liberation theology–inspired priests, and they have developed a strong analysis of the liberal state in Latin America in its articulations through class, racial, and gendered differences that are also ideological. Others participated in short-term workshops with agroecologists from Europe and Brazil who visited Padre Alcides's community outreach work in Puerto Caicedo during the 1980s and '90s. In the 1980s, Marxists were generally not environmentalists, and Heraldo is also critical of his time as a student, who he was then, and what he learned of the agricultural sciences at the university. Processes of unlearning and relearning dominant knowledge paradigms are complex and have varied trajectories and temporalities. Alternative agricultural practitioners such as Álvaro, Heraldo, Nelso, and Elva engage in sophisticated conceptual work that takes inspiration from Marxist-inflected Latin American political economy and agroecology, and what we might call affect theory, indigenous thought, and postdevelopment transitional discourses and initiatives.

When I began to write about my fieldwork, Heraldo and I engaged in various negotiations over the politics of ethnographic conceptualization. For example, the first time I shared with him an academic conference paper, he was quick to point out that agroecology was the wrong terminology to describe the diverse practices we had been witnessing. Rural communities are not "professional ecologists," but rather practitioners learning in and with a place. Neither was organic agriculture an appropriate description because organic agriculture can be reliant on commercial inputs. It can also sound "pretty and romantic," he said, without conveying the political stakes of breaking with dominant knowledge paradigms and defending other practices that transform not only the socioecological relations, but also the knowledge politics in a territory. These were opportunities to learn not to bundle up or capture things too quickly in concepts that were most readily familiar to me. We also negotiated the aesthetics of ethnographic writing. After reading one of my dissertation chapters on decomposition, he remarked how beautiful I made shit, urine, and manure sound. The poetics did not bother him. On the contrary, it was important to convey the sensorial and affective quality of the resonating farms. One thing was for me to transmit the conceptual work and proposals of rural communities, and another was my creative license to work with the concepts that I learned "in the field." The ideas that campesinos shared with me were not meant to be privatized. Heraldo said that just as he and other rural practitioners borrowed ideas from textbooks, scientific articles, public statements issued by social movements, local stories, and exchanges with neighbors and other rural communities, the concepts that I was learning to think with should also travel and should take on a new life through my analytical work and textual iterations of them.

Rural families embarking on what I conceptualize as *trajectories of selva apprenticeship* are becoming what Heraldo and the Regional Working Group of Social Movements (MEROS) call "Amazonian men and women." Amazonian men and women are becoming *selvacinos/as* instead of being campesinos/as, which is to say that they are learning to cultivate and be cultivated by the selva rather than clearing forest to work an open field, pasture, or *campo*. Instead of colonizing the selva, selvacinos/as are learning to assist the selva in recolonizing farms. This occurs at the same time that la selva remains relatively indifferent to human activities and desires, or as Heraldo and Nelso argue, *"La región es"* (the region is [and will continue to be]) regardless of state development plans that have reclassified tracts of the Amazon and its inhabitants as zones prioritized for industrial extractivism. More than an environmental subjectivity, I conceive of selvacinos/as or becoming Amazonian human to be an emerging relationality, an assemblage of practices that problematizes modern object–subject dualisms relying on an ontological division between "nature" and "culture," "labor" and "rest," and "working farm" and "natural forest." There is no campo to be improved, modernized, worked over, or extracted from. Nor is there a human with absolute agentive and sensorial mastery. The Amazonian human is one in whom the human—what it means to be human—is necessarily composed of and decomposing back into selva when one follows "her" rather than attempting to domesticate and colonize a savage, worthless, or empty landscape. Becoming selvacino/a cannot be separated from this assemblage of practices and their corresponding life philosophies. Similar to Marisol de la Cadena's discussion of "being-in-ayllu" in the Peruvian Andes, becoming selvacino/a does not refer to persons from a place, but the relational workings through which one becomes of a place (2015a, 102)—or as I see it, the ongoing trajectories of apprenticeship through which one learns to be composed by and decompose back into a place.[10]

UN SANCOCHO DE IDEAS PROPIAS DE AQUÍ

[A Stew of Ideas from Here]

Heraldo wrote his first propositional reflection on becoming Amazonian men and women in 1993. This occurred after he listened to a series of scientific experts (i.e., Amazonólogos) who had been invited to Putumayo by the regional environmental authority, now known as CORPOAMAZONIA, to discuss sustainable development. When Heraldo began working on his own concepts and promoting community spaces to discuss issues informed by, but significantly departing from, what he observed at CORPOAMAZONIA, local communities began to also refer to him as an Amazonólogo. Thus arose his need to clearly distinguish Amazonólogos from Amazonian men and women. As I mentioned when introducing the concept of lecturaleza, one way that Heraldo contrasts scientific practices from Amazonian ones is by differentiating between the extractivist objectives of "cosechar conocimiento" (harvesting knowledge) and participating in the creation of "conocimiento vivo" (living knowledge). Living knowledge is mutually constitutive of the daily labor of *cosechar comida* (harvesting food). Heraldo once told me: "I am not a scholar of the Amazon. I am an inhabitant, a producer with a technical degree. I am not a researcher or scientist with a doctorate. I do not seek to differentiate myself. I am different. I gain information through practices in concrete situations that more or less approximate scientific knowledge. I experiment to grow things and I go on learning. I may 'know' less about the territory, but I struggle for the territory with my neighbors because like them, I am of the territory. We are practitioners with ancestral and popular wisdom." In all my travels with Heraldo, I never heard him introduced to communities as an expert, but instead as a *conversador con la naturaleza* (someone in conversation or a conversationalist with nature).

I am not suggesting that Heraldo and the other families with whom I spent time categorically reject the techno-scientific knowledge produced by Amazonólogos. They consider specific scientific practices to possess subversive potential, such as those of agroecology. Subversive scientific practices are those that do not render rural communities dependent on synthetic or organic commercial inputs or services and technologies that can be creatively made on the farm using local tools and resources. They are scientifically informed practices that support the autonomy of rural communities. I think of this combination of popular practices with the lessons learned from "subversive sciences" as a mode of *alter-teching* that does not reject technological innovation, but instead subverts conventional hierarchical relations between "low" and "high" tech by first attempting to make what one needs. This is a situated practice of deciding what matters. Even when they find ecological notions necessary and helpful, the first priority of these families is to valorize and multiply the experiences of campesino, indigenous, and Afro-descendant communities. Scientific theories become important when they aid in reinforcing this intercultural priority, and when they acknowledge that modern dualisms between "nature" and "culture" were never a universal phenomenon, but rather a specific knowledge tradition and ideological project that did not succeed in eliminating other relational world-making practices and modes of existence.[11] Even the science of ecology, Heraldo once told me, does not function without local practices, seasonal solar and lunar cycles, prayer chants, good thoughts, and lecturaleza. "Science," he said, "provides knowledge, but we also follow the knowledge of the universe." For these alternative agricultural practitioners, modern sciences enable access to a partial reality among other realities. This said, agricultural sciences are seen to be a powerfully imposing tool often hitched to capitalist interests in ways that other powerful saberes (wisdoms) are not, and thus interactions with the former require a greater level of skepticism and precaution.

Neither is becoming hombres y mujeres amazónicos in opposition to lo andino (Andean-based logics and ways of life). The majority of Putumayo's inhabitants and their ancestors arrived from neighboring Andean states, and they continue to eat and cultivate a lo andino. From rain and water cycles to the origins of rivers that periodically flood and fertilize the Amazonian plains, the geologic histories of the Andes and the Amazon are inextricably linked. When tectonic plates began to edge toward one another, and the resistance and repose of solid, liquid, and molten ash birthed the Andean mountain chains around 65 million years ago, what would later be called the Amazon Basin formed from the resulting sediment, floods, and volcanic residue. This process was so tumultuous that the great Amazonian rivers and subterranean waters channeling into these fierce currents suddenly changed direction.[12] Previously flowing east to west into the Pacific, they now flow west to east toward the Atlantic Ocean. I was told that this molten transformation left behind invisible and magical tunnels that continue to connect the Amazonian plains and Andean foothills, and that allow spirit energies and people to change form and pass through these corridors. At all levels, frictive, formative, and sedimenting processes in the Andes directly influence(d) the creation of Amazonian plains, hills, valleys, rivers, and swamps with their corresponding diversity of soils that, as I have discussed, the homogenizing notion of "poor soils" serves to eclipse.[13]

Pensamiento propio as a sancocho is instructive because a stew is not a melting pot that claims multicultural tolerance while blending difference into a homogeneous soup that mysteriously always looks and tastes the same. Nor is a stew a soothing and benign broth. A sancocho is a popular dish that may be full of scraps and leftovers, laden with changing ingredients and regional delicacies. It has no set recipe even though it is a staple food. Different techniques go into its making. Sancochos must be prepared in stages due to the cooking time differential of their various ingredients, and thus they require a level of patience. Their distinct components stay chunky. They do not coalesce, but heartily "hang together" (Mol 2016), and there is really no way of preparing a small amount, so there is always some to share and eat again as the flavor modifies.[14] A sancocho is a way of imagining how to engage in intercultural dialogues that do not collapse differences or necessarily eliminate the generative aspects of conflict. Intercultural dialogues took on renewed importance in Putumayo and around the country after the 2013 National Agrarian, Ethnic, and Popular Strike, when summits (*cumbres*) were organized to create a working space for indigenous, Afro-descendant, and campesino communities to discuss their territorial conditions in the twenty states that had participated in the month-long protest.

Heraldo thinks seriously with Marx's use of the concept of contradiction, and it informs his analytical and political visions. He once told me:

> I am Catholic, but this does not prevent me from singing with the evangelicals or taking yagé. Siona people can learn from unions, campesinos can learn from the Siona, unions can learn from certain ecologists, ecologists can learn from Afro-Colombian communities. There are differences and contradictions that can coexist and others that cannot. One can be in solidarity with those who are different. We nonenemies can look each other in the face. There are *diferencias de manejo* [different ways of doing things], and there are *diferencias de fondo* [differences in the substances/essences of things]. We all maintain secrets. We do not have to share everything. Besides, they kill people in this country for talking too much. What we are doing is making *un sancocho de ideas propias de aquí*.

Heraldo and Nelso also maintain the ontological differences between the material constitution and philosophical underpinnings of, on the one hand, campesino gardens and integral farms and, on the other hand, indigenous ancestral cultivation areas or chagras. Chagras are a product of millennial thinking or, as Heraldo says, "cosmovision," constituted by a set of relations and beings more akin to what Eduardo Kohn (2013) describes as an "ecology of selves" in that the constitution of the chagra implies an awareness of the selfhood of the many beings that people the cosmos in which they live.[15] Farmers' integral gardens tend to be more informed by biological and ecological thinking, and in many cases entail an un- and relearning of relations in which the modern dualisms of "nature" and "culture" and "humans" and "nonhumans" have to be reexamined and deconstructed.

TENACES

[Tenacious Ones]

If Heraldo contrasts Amazonian men and women with what he ambiguously calls Amazonólogos, he also marks a political distinction between social leaders and what he refers to as *tenaces* (tenacious ones). Social leaders refer to a generation of regional activists who have reorganized since 2006, and who lead protests, denounce human rights violations, and negotiate with government officials to demand the recognition of rural communities as legitimate political interlocutors. If they have time to carry on farming or even live in rural veredas, many social leaders continue to be coca growers or have replaced coca with other acid-resistant monocrops. While tenaces stand in solidarity with these social leaders and the ideological struggles they represent and provide regional social organizations with the kind of alternative and solidarity-based technical assistance I have been describing throughout the book, they do not necessarily have the time or disposition to attend every protest, sit through bureaucratic meetings, or negotiate with elected officials. This is, in part, because their lives are deeply enmeshed with, and hence mutually obligated by, other beings and elements with whom they co-labor in gardens, orchards, and forests.

The successional, regenerative, and decompositional temporalities and everyday acts of care and labor involved in sustaining these relations are largely imperceptible within the official time of politics—progressive, conservative, or otherwise. In other words, the work of tenaces is not legible in the government reports on competitive agricultural yields or the four-year development plans that accompany elected officials. Nor does it correspond with bureaucratic office hours or news reports on popular strikes and uprisings. One can argue that tenaces primarily situate their alliance-building within the relations and conocimiento vivo (living knowledge) that form part of daily life and labor on farms, and relatively distance themselves from conventional political spheres and representational procedures because they do not want to be counted as having participated in state- or corporate-convened meetings. Mere presence on the attendance lists of such meetings, for example, fulfills the prior consultation requirement (*consulta previa*) that oil and mining companies are legally bound to engage in with ethnic groups and other collectively owned territories before engaging in legislative, administrative, economic, or infrastructural projects in these territories.[16] This is why with Heraldo, Nelso, and others we at times hiked for seven hours to remote areas because, as Heraldo explained: "Tenaces are dispersed men and women doing things without technical support. You have to look for them on farms and not in political spaces. They have other seeds and plants, communal intellectual property instead of knowledge that is monopolized, captured, and applied. Tenaces are not beneficiaries, but actors with local talent." As Heraldo and MEROS members see it, social leaders and tenaces are political allies who can draw upon their divergent strengths and shared experiences of struggle. However, social leaders are a different kind of tenaz than those primarily positioning themselves within the processes dedicated to cultivating farms as more than farms.

Yet over time it was hard for me to identify who might actually be a tenaz. In fact, I was never sure if I ever met one. Even when Heraldo would refer to himself in this way, it was never to suggest that he was exceptional or the only tenaz in the region. Furthermore, as I mentioned, he had occupied political positions in his life, such as twice accepting the position of provincial secretary of agriculture. Nelso had also run for office as a municipal deputy. There is no purism in their political engagements, much in the same way that they interact with certain scientific practices depending on the shifting potential for alliance-building. By the end of my long-term fieldwork, I began to press Heraldo to identify which of the different campesinos and rural families we had met might be tenaces. He always answered with vague replies. "It is hard to know," he would say. "There are different levels of tenacity. It is necessary to wait and see." I came to realize that perhaps tenaces are uncountable and unrepresentable, a nonnumeric placeholder for all those dispersed agro-life proposals that do not necessarily desire to become more—in other words, not the dream of a Putumayo replete with tenaces in the way that selvacinos/as or Amazonian men and women might come to inhabit a growing network of resonating farms.

It seemed to me that as a concept, tenaces allowed Heraldo to make analytical and political moves that *oppose* dominant agricultural practices and extractive models while at the same time *connecting* with other alternative practitioners and proposals. Tenaces might be akin to what Deleuze and Guattari (1996) called "conceptual personae," in that they play a crucial part in Heraldo's creation of other selva concepts. For Deleuze and Guattari, a philosophical enunciation requires movement by thinking it through the intermediary of a conceptual persona that is a living embodiment or illustration of the philosopher's key ideas. They describe similarities between the fictitious characters of novels and the concepts running through philosophical works. Tenaces signal Heraldo's movement of thinking that differentiates itself and is differentiated from other thought, not only in an affirmative sense, but also in the moments when his proposals are written off as utopian, or more dangerously, as radically autochthonous and communist. The concept of becoming a tenaz has come to be associated with Heraldo among the alternative agricultural practitioners and networks with whom he has made contact. It seems to me that tenaz signals an ontological shift in thought, action, and affect when one learns to follow the selva, a kind of conceptual intermediary that creates the conditions of possibility for a range of selva-life imaginings, material and ethical transformations, personhoods, and dreamings.

I am led to return to Heraldo's story of the Andakies, who continue to resist colonization by rendering themselves imperceptible and transforming into something else—into other forms of matter and nonmaterialities. He told me that the Andakies continue to exist unperceived among us, only at times appearing as pumas-aucas or in *pintas* (visions induced by yagé) to nurture the survival of different forms of selva life. "*Eran tenaces, siguen siendo tenaces. Son los antiguos. Son los más sabios que no se dejaron dominar. Conservaron su poder. Los tenaces siguen presentes. Seguimos,*" he said. (They were the tenacious ones. They are still tenacious. They are the oldest, the wisest ones that never allowed themselves to be dominated. They conserved their power. The tenacious ones remain present. We carry on.) Tenaces may conceptually and energetically accompany all those rural communities that are asking how to defend, reclaim, and cultivate alternative selva agro-life processes. They may be a force that inspires people in ways that neither coca crops and their official alternatives nor public policies have done before: to learn to work, eat, and shit *with* rather than against the selva.

GARDEN DREAMING

On two different occasions, I accompanied Heraldo to the Environmental Clinic in Sucumbíos, Ecuador, a permanent selva campsite and observation center built by the nongovernmental environmental organization Acción Ecológica, based in Quito. The NGO suffered persecution in recent years under former President Correa's administration due to their oppositional anti-extractivist stance, environmental defense work, and support of community organizing. On our second trip in 2010, Adolfo, a Spanish biologist and medical doctor and cofounder of the NGO, greeted us in a T-shirt that reiterated a refrain first heard in Ogoni territory in Nigeria over twenty years ago: "Keep the oil in the soil." It went on to read, "keep the coal in the hole, the tar sand in the land, the gas below the grass." Heraldo and I traveled from Mocoa to spend the weekend at a seed and food fair with local campesino and indigenous communities, and to learn about the legal, practical, and philosophical differences between food security and sovereignty at a popular education workshop where

FIGURE 5.6 Learning composting techniques at the Environmental Clinic in Sucumbíos, Ecuador, October 2010. Photograph by author.

Heraldo had been invited to share his experiences on the topic. At the time, the Ecuadorian military was under pressure from the Colombian government to conduct joint operations against the FARC-EP, who were known to cross over into Ecuador to buy supplies and weapons, and to set up base camps and engage in R&R.[17] Increased militarization along the San Miguel River, which serves as the border between the two countries, had unsurprisingly led to increased human rights violations against local communities. The Ecuadorian government had also begun to refuse land titles to Colombian immigrants and refugees, discriminating against them out of fear that they were involved in illegal activities or collaborating with armed groups. At the start of the workshop, community members reported that more sources of local drinking water had been impacted by increased seismic exploration on the part of the oil industry. Many residents had taken to placing signs in front of their houses that read: "This farm has been contaminated by such and such multinational oil company."

It was a long day after our travels. We went to bed early because at five in the morning a group of children would make the rounds to call people to the campfire to participate in a ritual to interpret dreams. Pablo, a Quechua chamán who had been asked to perform the ritual, woke up an hour before everyone else to prepare the *guayusa* tea leaves. By sunrise, people were loosened up and warmed by the caffeinated tea, and the dreams began to flow. Pablo explained to us that there are direct and indirect dreams and instructional and preventive ones. When it was Heraldo's turn to speak, he began by sharing a recurrent dream in which he is prepared to leave for a trip but ends up missing the bus because he is busy doing errands. Pablo explained that this is most likely because something is blocking his learning process and personal projects, and that he has to catch up with the velocity of the bus. Heraldo then mentioned another recurring dream in which two snakes are battling each other. Pablo said that if the snakes or dogs do not bite in the dream then this indicates a pending conversation or discussion, but not a dangerous or deleterious conflict that Heraldo needed to be worried about. The recounting of these two dreams seemed to be leading up to a more perplexing issue. After a brief pause, Heraldo went on to share that he often feels an accompanying feminine presence when he is in a half-awake state, and that after a day of work on the farm he lies in bed, but often continues to see plants, vegetation, and gardens floating in front of him. Sometimes there is a woman walking around the garden in a white robe. There are so many plants, plants that he has never seen before, and fruits and vegetables that he has never tasted. At other times, when he is about to drift off to sleep, he hears a violin or guitar playing in the distance. He wondered whether these could be good or bad *duendes* (imps, magical creatures) that are known to lure people into the selva with melodies and sweets for either romantic or prankish purposes. He asked the chamán if these might be hallucinations, and if he should be concerned.

Pablo first inquired whether he spent much time alone or if he was married or lived with a woman. Heraldo explained that he had married only three years earlier, and that his wife did not live permanently on the farm or accompany him in much of his work. Pablo's assessment was immediate: loneliness and untapped potential. La selva was directly appearing before him and was attempting to seduce him. The woman in the white robe was urging him to do more. He can do more. In fact, he should do more. Pablo described other cases that have occurred in the forest, on the banks of rivers, and in nearby towns where men have been hypnotized by the selva and never return from a hike or a swim. Revealing only slight concern, Heraldo asked what happens when the selva seduces someone and draws them away. "No, nothing. Everything is fine. One lives there. It is like this reality, but it is different," Pablo replied. Heraldo seemed at ease with this response, and it was only during our trip back to Putumayo that I had the opportunity to ask him about the chamán's interpretation. I began with a joke. Perhaps Heraldo had already crossed over to the other selva plane of existence, I said. He laughed, but I knew my joke to be ridiculous. We were of course not on the other side, at least not yet. The other side would be like this, but different, a different Putumayo, different Putumayo(s).

When we got to talk more about the act of dreaming, I learned that Heraldo considers dreaming to be better than reading fiction. "I prefer dreaming to reading a novel," he told me. This was because his dreams inspired him to actualize different gardens and orchards. They energized him to seek out a greater diversity of seeds and encouraged him to carry on with the construction of the integral farm school, to keep on cultivating his "fifth attempt." The fruits and vegetables that floated before his eyes when he was bleary-eyed and oscillating between a half-awake and a dream state were a different twist on the popular phrase *soñar despierto*, which is usually translated into English as daydreaming. With farmers and various friends in Putumayo, we had conversations about soñar despierto. This is a mode of dreaming that implies having one's feet touching the ground, learning where one is standing in terms of becoming attuned to one's surroundings and the potentiality to actualize material change, to actualize the very substances of one's dreams.

The woman in the robe, the feminine selva presence in Heraldo's garden dreams, may not necessarily be the seductive existence of a female lover. Nor, of course, is seduction only a female capacity to begin with. If it is a romantic pursuit, this is a laborious, risky, enredando (entangling), transformative process that entails undoing and redoing, unlearning and relearning. It is a seductive presence that obliges shifting affects and also inspires recalcitrance. It was on this trip to the Environmental Clinic that I began to learn how important dreaming is in processes of cultivating eyes for her—cultivating attentiveness to the selva, respect for her indifference, vulnerability, and endurance, and for the generative power of garden dreams and vegetal presences. To dream profoundly, Gaston Bachelard wrote, is to dream matter. "Dreams come before contemplation. Before becoming a conscious site, every landscape is an oneiric experience. . . . But the oneiric landscape is not a frame that is filled up with impressions; it is providing substance" (1994, 4). Behind or beneath the imagination of forms so often privileged in discussions of cognition and aesthetics there lies, according to Bachelard, an imagination of substances. Material projects in the process of execution have different temporal structures than their intellectual projects. We might call this temporal structure a dream time that is less a faculty of an individuated subject, enabling one to act upon and transform the world, and more of a relational becoming involving a multiplicity of elements, beings, and materialities—a woman in a white robe walking through a never-before-seen garden, the vibrating resonance of creeping

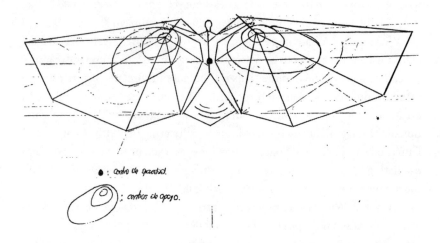

: Centro de gravedad.

: centros de apoyo.

FIGURE 5.7 Heraldo's design for the *Murciélago Humano* (Vallejo 1993a).

plants, and the appearance and disappearance of footprints of pumas-aucas that appear when called upon to share selva wisdom and ancestral stories.

THE HUMAN BAT

On one occasion while I was perusing Heraldo's bookcase at the farm, I found myself staring at an inconspicuous handwritten document that had been tucked between stacks of papers. It was stained, creased at the corners, and held together by a rusty and embedded staple. For some reason, I started reading it from back to front. The last three pages contained drawings of what looked like a human body with mammoth outstretched wings extending from arms, torso, and legs. One page included dimensions for the fitted members of the wingspan, and another, circular drawings indicating points of gravity and muscular support. I quickly flipped to the cover page and read the title: *El Murciélago Humano* (The Human Bat), written in 1993 when Heraldo lived in Villagarzón (1993a).

The document opens and closes on a cautious note. Heraldo prefers not to present his proposal to the public until what he refers to as the "dream of flying" becomes a reality. Otherwise, he writes, he might be labeled *chiflado* (nuts, screwy). Clandestine accomplices are needed, and anyone attempting to take flight should be well versed in a few important safety recommendations:

(1) only short and low flights should be conducted if the human bat is not in excellent physical condition; (2) it would be wise to travel with a parachute in case of unexpected trouble in the sky, although a parachute will also contribute additional weight; and (3) a helmet is advisable to protect against cold winds and insects that might obstruct the human bat's vision. Heraldo writes that he was inspired to design the murciélago humano by his admiration for the acrobatic and panoramic capacities of birds. He dreams of achieving a different vision and mode of transport across the landscape, sailing high above trees and foothills much like scuba divers glide through the depths of the sea.

Making mention of hang gliders and skydivers, Heraldo outlines the problem of their liftoff from the ground where air currents do not move in an upward direction and the wind exerts no force to aid against gravity. I am not sure how he knew—if by reading or observing or both—but he writes that the aerodynamic mechanics of hens and eagles are too complex to follow. Bats, however, have a particular sort of forelimb constituting webbed wings. In place of flapping their entire forelimbs like birds, they flap only their spread-out digits covered with a thin membrane uniting the forelimbs to their abdomen. While the weight of bat bodies and their wingspans are not naturally balanced, a strong muscular development produces the force necessary not only to take off, but also to sustain flight. This makes bats the only mammals naturally capable of such a feat. According to Heraldo, it is possible for the human bat to exhibit similar muscular potential. As his diagram indicates, arms/wings must be unrestrained in order to produce enough muscular force by moving simultaneously up and down and forward and backward as if treading water. Heraldo provides instructions for liftoff and landing, four steps in the motion of the wings, the materials needed to construct them, and how to place tiny bags of gas to modify aviators' different body weights. In all honesty, he writes, he cannot understand why we are not accustomed to the sight of more human bats flying in the sky. "Why is it that we always wait for others to make the things we need?" Bat wings, bat membranes, human muscle, human bat, flying device, flying-Amazonian-bat-human-contraption. The only thing to do is to experiment and imagine, suggests Heraldo. Just by reading this story perhaps we become a covert accomplice in Heraldo's flying human bat dream.

I do not share this dream because I witnessed him soar high above the rooftop of his farmhouse and off into the distance of the Andean-Amazonian foothills. By the time I met Heraldo, he had temporarily shelved the idea because he was busy working toward actualizing many others. However, I love and always come back to *El Murciélago Humano* because it says so much about Heraldo and many of the rural families and alternative agrarian collectives that I have met and accompanied over the last decade—their experimental curiosity and conceptual playfulness; their ability to jump ahead of themselves, ahead of the way situations are presented to them as the limits of possibility or the crazed stirrings of a chiflado. In this proposal, Heraldo's dream is to achieve a different mode of transport and itinerant vision, or ojos para la selva. He expresses an urgency to make what one needs and to regain a level of autonomy that has been weakened or systematically dismantled. This autonomy might be reclaimed by enlisting accomplices who are able to trust in their own capacities and who are open to following the flight of bats as dusk hovers over the piedemonte without certainty about the outcome of the journey. Among other scholars, feminist essayist Rebecca Solnit has taught us to see optimism and pessimism as two sides of the same coin: both try to remove uncertainty from the world (Montgomery and Bergman 2017). I understand the human bat to be a creative proposition that questions fixed ways of relating, which, in turn, destroy one's capacity to be responsive and inventive, to imagine and dream differently without the security of guarantees.

6

WHICH SOILS?
WHERE SOILS?
WHY SOILS?

HERALDO: *I see the farms as systems that capture energy. Plantas [plants] are the sensors.*
CORPOAMAZONIA TECHNICIAN: *Heraldo, how many plantas [generators] do you intend to set up?*
HERALDO: *No, hombre. I am talking about plantas [plants].*
CORPOAMAZONIA TECHNICIAN: *Heraldo, plantas [generators] are not cheap!*
HERALDO: *Hombre, I am talking about arbustos [bushes], matas [shrubs], and plantas [plants]!*
CORPOAMAZONIA TECHNICIAN: *[silence]*
HERALDO: *I am afraid you simply do not understand.*

THAT WHICH DID NOT BECOME SOIL

How do different ways of relating to soils—including that which may or may not be enacted as a scientific object called "soil"—produce differential economic, political, and environmental policies, legal structures, technological innovations, and forms of territorial ordinance? Put another way: What kinds of potentialities germinate from different sets of human–soil relations, relations that may unravel or destabilize concepts of the human and of soil and of their hyphenated pairing? These questions and their stakes are always profoundly situated at the same time that they are critically planetary. Rather than assuming a singular planetary scale or a singular world, in this book I have attempted to articulate the worlds that are enacted in situated relationships with soils as well as with those relations that do not become "soil."[1] For example, if industrialized and chemical conceptions and treatments of soil emerge from a history that is ecologically bound up with the violence and depletion that we currently call anthropogenic climate change and a geologic epoch referred to as the Anthropocene, then these soils—chemical and industrialized ones—form part of the very conditions that made what we know as the Anthropocene possible. The stable, meter-deep, mineral-rich, fertile soils that have been relied upon for limitless resource extraction and economic growth—harvest after harvest of calories and profits at the hands of chemical input substitution and mechanization—were always a specifically situated and never universalizable soil. It is this soil, precisely the recognition of the limits to assumptions about an unchanging standing reserve entity, that concerns over anthropogenic climate change aim to find techno-scientific and geoengineered solutions to rescue and fix.

This may be the "rediscovery" of the ecological limits, fragility, and corporality of soils that has been alarmingly signaled by campaigns scaling from the Food and Agriculture Organization of the United Nations' (FAO) 2015 declaration of the International Year of Soils to the Geographic Institute Agustín Codazzi's 2009 campaign, The Year of Soils in Colombia. Dorion Sagan (2011) calls this paradoxical phenomenon the "return of the repressed," referring to the way that the blockage and exclusion of all systems, living and nonliving, that render human life possible, has hauntingly returned to stress and press upon a whole range of social and natural sciences. In other words, it is not that a stable soil has been disturbed and provoked to crisis, but rather that capitalist agriculture has increasingly stabilized a specific idea and treatment of soil to a mechanized and rigid timetable of production that has generated volatile changes, destruction, and loss.[2] It is this destruction that has now come back to haunt the heterogeneous field of soil science and the modern agricultural sciences more broadly.

Whenever one hears alarming statistics and reads panicked headlines about the accelerating degradation of the "world's soils," I suggest that it is important to ask what the world's soils occlude. I propose this in the spirit of the "anthropos-not-seen" (de la Cadena 2015b)—or those ways of making and doing life that have been disappeared and marginalized not only by, as Natasha Myers (2019) puts it, "multitudinous instantiations of colonialism and capitalism," but also by "a singular optics of conventional Anthropocene thinking" that uncritically assumes a blanket concept of humanity, history, and the geological record. In this book, the not-seen includes what are considered agriculturally "unproductive" soils whose particularities are actively stigmatized or overlooked, as well as the ongoing histories of dispossession and criminalization that sever people from the relational entanglements that make selva-soil, gardens, farms, and territories. Taking a step back even further, this book asks where soils are being enacted as stable, albeit evolving, objects, and where soils do not emerge as such because there is no entity that can be separated out from a relational continuum of human and other than human life.

It was the campesino families, agrarian movements, and alternative agricultural networks and practitioners I accompanied throughout the Andean-Amazonian foothills and plains that taught me to ask what we can learn from "soils" that never became industrialized or chemical: all those materialities and temporalities that undermine and escape dominant taxonomic classification; that have always been transitional states rather than stable entities; and that resemble litter layers that decompositionally retire instead of lending themselves to relentless growth-oriented production. What assemblages of practices are attached to these soils or nonsoils? What worlds are these practices attempting to reclaim, imagine, cultivate, and defend? Rural communities in the western Amazon are repositioning the urgency to learn "where they are standing" in struggles to recover what is always a relative autonomy from capitalist agriculture, extractive industries, and a militarized, development-oriented state apparatus. Within the particular agroecological conditions of the region, learning where one is standing entails engaging in a radical shift from searching for a "quality," mineral-rich evolving parent rock or meter-deep stable ground to learning how to cultivate and be cultivated by selva-soil cycles that necessarily keep turning over, aerating, regrowing, and rotting.

POSSIBILITY AND CONSTRAINT IN THE RECLAIMING OF PRACTICES

The heterogeneous group of state soil scientists and technocrats I accompanied and interviewed in Bogotá and other urban areas of the country often attributed soil management and planning debacles to the lack of a coordinated vision between the Ministries of Agriculture, Environment, and Mining and Energy; the failure to consult soil scientists in the design of natural resource legislation; the soil's lack of "natural" charisma; and corruptive loopholes that allow soils-as-territorial-units to be employed in activities other than their appropriate and designated vocations. Inspired by Stengers's commitment to supporting the reclaiming processes of scientific practitioners, and their partial manner of relating to the world, this book also attends to the specific reclaiming practices of a heterogenous grouping of soil scientists. By "reclaiming," Stengers (2017) pairs struggles against the rendering of all life, ways of doing things, and manners of existence as a potential resource with the need to heal from the devastation of what makes scientists think and imagine—their institutional symbiosis with state and capitalism (393). I engaged with state soil scientists who are attempting to address the ways in which what they call soil has been reduced to a singular economic vocation at the same time that they confront humanitarian pressures to enlist soils to respond to population growth, expanding urbanization, global crisis in food production and distribution, and widening gaps in poverty. In addition to this growing list are what environmental management frameworks refer to as ecosystem services and the myriad terrestrial life-sustaining roles and functions that soils are expected and asked to perform. Soil scientists are trying to build spaces of articulation between scientific practitioners within a disciplinary legacy that prioritized chemistry and physics over biology and microbiology at the expense of historically hitching their discipline to the imperatives of industrial agriculture. They are also reexamining their role in producing what they characterize as the soil's chronic political and legal "anonymity" in a society where soils as environmental relations and natural bodies have overwhelmingly not come to matter. During the IGAC's Year of Soils events, I listened to debates over how an ontologically complex object had come to "fail" its specialists and spokespersons, and how specialists had been unable to protect their object of study and affection. The institutional momentum gained by the campaign led to the design of a national Policy for the Integral Environmental Management of the Soil (GIAS) that coincided with the FAO's support to initiate a South American Alliance for Soils in 2015.

What remains unclear is how these state soil scientists will mobilize a concept of "soil health" that considers corporal vulnerability at the same time that the nation's development model depends on intensified resource extraction and privatization. Furthermore, while soil health makes a direct connection to human and environmental health, the concept of health may become another mechanism for infinite economic growth that fails to liberate soils, scientists, or rural communities from exploitative capitalist relations and industry-funded research mandates. Colombia's then fifty-year-plus social and armed conflict, ongoing territorial conflicts, enduring unequal patterns of land distribution, and systematic violence against social leaders and defenders of land and territory were rarely topics of discussion during the Year of Soils campaign. For this reason, the title of the 2016 National Soil Science Conference caught my attention: "Healthy and Productive Soils for Peace in Colombia."

In a country where agrarian scholars and activists have long argued that *"para sembrar la paz hay que aflojar la tierra"* (to sow peace it is necessary to loosen up the land) (Fajardo Montaña 2002), the links between soils and peace have been far less examined, and, as this book elucidates, extend far beyond the important work of humanitarian demining in the country's post-peace accord, transitional justice scenario. The socio-ecological transformations necessary for the building of a lasting and inclusive peace create renewed opportunities for the country's agrarian and ecological sciences. If many rural communities and agrarian-based popular movements are engaging in a *territorial opening* that is changing their relationships to a living world below, then what kinds of openings might soil scientists engage in as they become skilled at cultivating broader cross-disciplinary alliances? How might a greater number of soil scientists and agricultural extentionists learn from selvacinos/as, alternative agricultural practitioners, and rural communities if the latter are not required to first perform the scientific equivalence of their practices and know-how? As this book demonstrates, these questions must be posed within the situated struggles and continuing destructions that soil scientists and rural communities respectively confront within the political-economic conditions wrought by a national "energy-mining locomotive" development model, intensified resource extraction, and agro-environmental conflicts across the hemisphere that may be perpetuated and backed by armed actors and criminal networks. These questions seem ever more pressing in the country's current political conjuncture, where the official demobilization of the FARC-EP has caused a power vacuum, allowing illegal mining and logging and extensive cattle ranching to expand in newly opened territories that were previously impenetrable, and hence largely conserved during times of war.

My initial assumptions about fixed locations of political subjugation and asymmetry between scientific and non- or not-only scientific practitioners were quickly turned on their head when I attended my first lab meeting in the agricultural microbiology laboratory of the National University's Biotechnology Institute in Bogotá. I listened to soil microbiologists explain the constraints placed upon their research and professional positionality due to their institutional affiliations, funding sources, and alliances with industrial guilds. Similarly, IGAC agrologists shared their frustrated attempts at producing a "minor science" able to create a line of flight from extractive and productivist logics when they conduct soil surveys, engage in taxonomic classification, and make land use recommendations for the country's Amazon. In both cases, these groups of scientists rendered explicit the links between their knowledge production and power. While the rural communities I accompanied also shared stories about the structural violence, economic and political dependencies, and dominant technical models that impinge on their forms of making and doing life, they did not characterize their positions as particularly marginalized. In part, this is because they situate their alliance-building within the ecological relations and conocimiento vivo (living knowledge) that forms part of daily life and labor on farms and among selva rather than focusing on state-centric representational spheres.

These rural communities taught me that in order to "decolonize their farms"—and, along with this, their kitchen tables, "medicine cabinets," taste buds, intestines, seed networks, and marketplaces—they first had to decolonize dominant techno-scientific, chemically conceived, and market-oriented notions of and relations to "soil." Beyond a simple inversion of analytical symmetry, they do this by making a series of counter-dualistic moves that create openings for tense and potentially collaborative relations between scientific and not-only scientific practices—what I was drawn to conceptualize as processes of alter-teching that do not simply reject technological innovation. Rather, these families subvert conventional hierarchical relations between "low" and "high" tech by first making what one needs and beginning to cultivate ojos para ella. This ethnography emerged from taking serious decolonizing enactments of asymmetry as a conceptual and political—or better yet, life proposal—that displaces the primacy of "knowing" in favor of ongoing processes of unlearning and relearning.

ON THE POLITICS OF MOLECULAR TRANSFORMATIONS

The philosopher Michel Serres (1995a) asks if there is a time before time, a time of origin when the contours of the physical universe were not yet thought to be fixed; a time before the humanly constructed timescale of calendars, chronicles, and clocks. He refers to this as the "basic" or "meteorological" time of multiplicities and elemental surgings, of the earth's molten upheavals, the chaos out of which intelligible form and meaning are forever emerging and by which they are continuously reabsorbed. Basic time is not to be conceived as prior to or superseded by the time of human history. It remains contemporaneous with historical time as its unacknowledged precondition, resurfacing in periods of crisis and transformation as an abiding reservoir of new possibilities. While Serres is inspired by the possibilities that latently subside within and emerge along with dramatic upheavals and surgings, the everyday practices I observed in the cultivation of "farms as more than farms" introduced me to the kinds of possibilities emanating from molecular processes of transformation—decomposing hojarasca; leafcutter ants dragging shreds of leaflets across the underbrush; the regrowth on the side of a roadway that for some is "weeds" and for others a salad. The multiple layers of meaning and materiality and the vital frequencies of the first Amazonian farm school I visited, La Hojarasca, engendered the very possibility for the "dream time" of this book as an intellectual project, which aspires to follow that which grows in the midst of glyphosate-exposed and criminalized life. There is a musical cadence to the word *hojarasca* as it rolls over a tongue—silent h and j, open vowels o and a—hojarasca; that which necessarily dies in the process of regenerating; a death that is distinct from acts of eradication that are justified by the aim to ensure regrowth. Gabriel García Márquez's (1972) short story collection with the same name, translated into English as *Leaf Storm*, starts *in medias res*—in the middle of things—to create a world that is transforming in order to go on existing after waves of violence have uprooted and rocked a place. In a similar vein, this ethnography has sought to follow emerging and dispersed agro-life processes, shifting affective charges, and multiplying vital resonances—life processes that struggle not to transcend, but instead to sink into the wounds of deforestation and degradation; to regeneratively retire and decompositionally recompose without the guarantees of enduring permanence or overarching success.

Political theorists in Latin America and elsewhere have reoriented our attention away from state-centric models for social change precisely because this focus may lead us to overlook diverse potentialities existing in the present (Baronnet, Mora Bayo, and Stahler-Sholk 2011; Gutiérrez 2006; Zibechi 2007). Along with potentialities, as this book describes, I have paid attention to the everyday material practices that contribute to the creative ways that socioecological transformation actually strives to come into existence and endure. Nishnaabeg scholar Leanne Simpson (2011) and Papadopoulos, Stephenson, and Tsianos (2008) respectively argue that these material practices have been obfuscated by twentieth-century political thinking's fixation on quintessential oppositional events, such as strikes, revolutions, political elections, and built-up social movements, which are generally never in the present, but become designated as events in retrospect or anticipated as future possibilities. Without ignoring collective human actors and macro-level power relations, or the conventional key events of political economy, attention to emergent, everyday material practices and ongoing acts of reclaiming and relaying may lead us to consider the delicate balance and qualitative difference between what Ghassan Hage (2012) characterizes as oppositional politics and alternative practices: crudely put, the necessary balance between politics aimed at contesting, resisting, and/or defeating an existing order—the "anti"—and the practices in the present aimed at providing alternative material conditions to this very order—the "alter."

This book follows altering states, the not dead, but dying and rotting layers of leaves and their relational vulnerabilities, cyclical process, and potentiating force. Sinking into the impermanence and intermediacy of hojarasca obliged me to rethink my ethnographic practice and to incorporate poetic and literary genres of writing that hold in tension and articulate the varied rhythms and intensities of geologic, human, microbial, and vegetal materialities and temporalities. In addition, the poetic interludes that appear within and at times interrupt the text address the acts and latent threat of violence that explode into and accompany the everyday and produce gray spaces between official times of war and transitional periods of peace. In an ethnographic sense, I am primarily focused on the cultivation of life in the midst of death and contamination; however, the conceptual and aesthetic work of poetry stays attuned to the ways in which violence is a foundational and destabilizing condition of daily life even when it was and is not the explicit focus of my interlocutors. Hojarasca and selva-soil interactions complicated my understandings of the possibilities for life and death and underlined the limitations of modernist biopolitical binaries between what is alive and what is dead, and what is a criminal life and a justified death under conditions of war and ongoing conflict.

In the rush to move forward toward a post-glyphosate political moment after the legal suspension of the use of glyphosate in counternarcotics operations in 2015, certain government officials in Colombia were quick to deem aerial fumigation and its ongoing deleterious public health, environmental, and humanitarian impacts a thing of the past. In part, the national government had begun to make an incremental shift toward integral antidrug approaches that hinged on replacing discourses of criminality with those of vulnerability. However, conventional conceptions of vulnerability may be what Isabelle Stengers and Philippe Pignarre (2011) call an "infernal alternative" that continues to treat coca growers and disputed rural frontier territories as stigmatized objects of intervention in the most reductionist political sense, as problems that must be solved. Such an approach continues to overlook coca growers, campesinos, and agrarian movements as legitimate political interlocutors participating in hard-fought struggles and ongoing popular processes to collectively determine the socio-ecological, political, economic, and transitional justice-making present(s) and future(s) of their territories. As this book contends, these struggles are not only over human rights and land rights, but also over the right to live and die well, to live and die differently, to avoid the infernal alternative of "extracting or being extracted from." Indeed, with the election of right-wing Center Democratic presidential candidate Iván Duque in June 2018, there is much uncertainty about the future of the country's peace accords given the blatant continuation of narco-trafficking and reemergence of paramilitary actors and ongoing murder of social leaders, human rights defenders, demobilized insurgents, and environmentalists. In addition, chemical warfare as a legitimate counternarcotics strategy is far from over given current geopolitical pressure from the United States to reinstate aerial fumigation of agrichemicals.

HANDS "RIDDLED" WITH FUNGUS

Thick fingers, nails cracked open. Dirt-caked knuckles inhabited by fungus, an extension of the hojarasca that Heraldo communes with. Riddled with fungus is what I saw first. Foolishly, I even gave him a pair of rubber gloves to use on the farm, which actually would have made the organisms proliferate in captivity. Heraldo accepted them with diplomacy, but I never saw the gloves again. The day I was almost bitten by a venomous spider he suggested, tongue in cheek, that perhaps I should wear them when working alongside him in the garden. Fungus perhaps, but in no way riddled; if his hands got infected he would know what to do with them. "These organisms are tenacious," he would often tell me. They did not budge, only at times appearing to retreat or fade into the contours of his flesh. However, they never extended much beyond his knuckles either. Some days they would act up, and he would think to spread a mustard-colored mineral over the sorest patches. These were the areas that he tried to cover with seeds or leaves when we took photographic inventories of the farm and he positioned different plant varieties in the palms of his hands for the camera.[3]

Mostly, Heraldo and the fungus seemed to have settled into a kind of tense partnership. After so many years, at least the decade I have known him, he had stopped playing "host." Human flesh, fungus, dirt, scars, scabs, and the like had become almost constitutive of each other. By this, I do not mean to suggest that they live in neutrality or harmony, but instead that they find themselves participating in a relationship akin to what Stengers and Pignarre (2011) call a "symbiotic event." By this they refer to "a matter of opportunity, of partial connection" that does not engender an encompassing unity or common project (often understood as having the same interests in common) but breaks the indifference between beings whose "diverging interests now need each other" (60). They suggest that these events may indicate a way out of the either/or trap that haunts us: either universality, meaning that all practices have something in common, or else relativism, meaning that each practice has its own incommensurable standpoint, and that practices are thus blindly indifferent to each other, except insofar as they destroy or are destroyed by each other. I came to think of Heraldo's hands as a kind of event where human flesh, fungus, and soil keep on diverging, even wounding and irritating each other, as they de-

FIGURE 6.1 Conducting a photographic inventory of the farm with Heraldo. Mocoa, Putumayo, January 2011. Photograph by author.

fine what matters to them. In the process, a kind of rapport is established that enables fungus-riddled hands to cultivate and defend modes of life that are claimed to be backward, and thus an obstacle to the guaranteed calories, production, and profits of industrialized agriculture.

I became more aware of the radical proposals at work in this skin-fungus-soils event after observing the battered hands of local *raspachines* (coca harvesters). The repetitive friction of scraping (*raspar*) coca leaves off of branches produces a qualitatively different kind of wound. Fingers are stained red and often wrapped with torn scraps of pillowcases or old T-shirts that are tied from the wrist and extend outward to produce what look like tattered mummy fingers. These pieces of cloth attempt to protect fingers from bleeding blisters and the irritating bites of leaf-eating larvae and prickly seeds that blow into coca shrubs. The ripped rawness and hard work of these hands were able to achieve a level of economic sustenance. However, these hands could not engender events that were able to produce relevant ways of resisting what had come to criminalize them—the biopolitical dream of a world morally and physically improved by coca plants, and thus cocaleros' and raspachines' necessary eradication.

FIGURE 6.2 Raspar coca. San Miguel, Putumayo, January 2014. Photograph by author.

PEACE, NOT POISON

It rains hard and heavy for short spurts in San Miguel, Putumayo. It sounds like a can of nails pelting against the zinc sheets of the roof. Not one, but a hundred cans of nails. It is the kind of rain that comes and goes, and when it comes no one tries to speak. It would be like shouting next to a waterfall, all hands and exaggerated gestures. There is no use in straining our voices, so instead Pedro, Yesenia, and I take refuge inside the house. We sit on the wooden bench that serves as chair and table, a tiny island in the hall. Wooden planks divide the kitchen and the bedroom. The rest of the house is made of cement. Yesenia tells me that at one point there was talk of moving the kitchen up to take better advantage of the natural light, but the project was left half-constructed. The wood has begun to rot from all the rain.

A dog named Cusco joins us today. He is named after the historic city in southern Peru. His owner said that the coca leaf would pay their way to witness the Incan ruins. The man planned to travel south with Cusco since the dog always accompanied him, walking two paces ahead or four paces behind. One dawn, this man crossed the river to Ecuador to do an errand for the guerrillas. When he was landless and decided to settle down with a wife, the Frente 48 granted him a parcel of land to sow his coca crops. Now he is dead, blown to pieces by a smart bomb after the United States sold the technology to the Colombian government, and the military conducted an operation on the other side of the border. Cusco survived, but people did not see him for months. Then one day he swam back across the river. An orphan. Now he wanders from farm to farm and stays on several nights anywhere they feed him *cucayo*, the leftover rice that hardens to the bottom of the pan. Today it is our turn. Cusco sleeps calmly on the porch until he hears the sound of helicopter switchblades slice open the sky. Firecrackers are the worst kind of torture. One crack, pop, bang, and whiz, and he returns to the morning of the bombing. Ears bent back, nose to the ground, beady eyes, frantic tail. There is nothing to do but bolt, run like the wind, run to the mountains, run, swim, pant, swim, run. People remember when Cusco had a round belly. Now when he returns from his flight, hair still on end, we find a ball of terror tucked under skin and bones.

During the rainstorm, Pedro, Yesenia, and I sit looking at the wooden boards dividing the rooms of the house. They tell stories. Stories carved with knives, drawn in marker, and charred with lighters and matches. Stories of when this farm was one of the largest cocaine kitchens along the border, stereotypical stories of planes landing on airstrips to carry off white gold. Children who remember the air raid that killed Cusco's owner had drawn other stories in bright red marker. A house pelted by machine gun fire. People, hens, turkeys, and dogs are all running for the hills. The wooden boards remind me of a mural in one of the last towns that one passes before beginning the trip downriver. It was painted by young people during the first years of the war on drugs to portray life before and after aerial fumigation. There is a decisive before and after: after dead rivers, dead animals, dead grass, dead plants. Vacant houses. Defoliated trees. Burnt soils. Skin irritations, inflamed eyes, bronchial conditions, birth defects, and dying cells. Children know this. Everyone knows this. Monsanto certainly does. Its Roundup-Ready product label provides a long list of hazards if the chemical is inhaled, swallowed, or enters into any other direct contact with humans and domestic animals. SEEK A POISON CON-TROL CENTER! SEEK FRESH AIR! IMMEDIATELY RINSE OUT EYES! DRINK MILK! Alongside the mural someone has spray-painted the words *Peace, not poison!*

When the rain calms, we go back outside to resume our conversation about what it means to give up coca, to start recovering creole seeds and reforesting the river banks, and to sow Amazonian fruit trees this far down in the "boot" of Colombia's national territory. It is not necessary to ask why or how these families ended up here cultivating coca crops. The dusk air quickly saturates with the intense hum of an orchestra of crickets. Every now and then we hear the high-pitched call of mochilero birds erupt out of nearby selva canopy. The low growl of howler monkeys maintains a steady beat. Yesenia's tone is at once cautious and hopeful: "*Seguimos con pequeños pasos. Lentos para que sean bien dados.*" (We continue with small steps. Slow so that they are well placed.) It is not a triumphant response. It is a moment like this: sunset from a hammock, a company of fireflies, the blink of their white light against a darkening night, a sky without the color show of a military air raid.

INTRODUCTION: LIFE IN THE MIDST OF POISON

1 For a comprehensive historical memory of the origins of aerial fumigation policy in Colombia, see the report published by MamaCoca and authored by María Mercedes Moreno (2015), http://www.mamacoca.org/docs_de_base /Fumigas/Memoria_historica_de_los_origenes_de_las_fumigaciones _MMMoreno_9mayo2015.html, accessed on May 6, 2017.

2 The mixture of glyphosate utilized in aerial spraying has been estimated to be 110 percent more concentrated than Monsanto's commercially available version called Roundup Ultra. Like most worldwide industrial applications of glyphosate, other chemicals accompanied the herbicide to enhance its activity and make it adhere to plants in a humid tropical climate: two surfactants, polyethoxylated tallow amine (POEA) and Cosmo Flux 411 (Vargas Meza 1999).

3 The Antinarcotics Directorate of the National Police provided these official statistics on August 18, 2015.

4 Despite the quantities of illicit crops sprayed, national coca-monitoring surveys have reported a steady increase in coca, with the number of hectares returning to and even surpassing the quantity that existed in 1999 when aerial fumigation policy first intensified (UNODC 2015). Resolution 006 was passed on May 29, 2015. See http://www.aida-americas.org/ngos-celebrate-suspension-of-aerial-spraying -advance-in-colombian-environmental-law, accessed on March 16, 2017.

5 A nice example of this is Juno Parreñas's discussion of methods in terms of multispecies ethnography. See https://aesengagement.wordpress.com/2015/09/15 /multispecies-ethnography-and-social-hierarchy/.

CHAPTER 1: FROM AERIAL SPACES TO LITTER LAYERS

1 For an in-depth ethnographic analysis of the US foreign policy-making process in the design, implementation, and assessment of Plan Colombia, see Tate (2015).

2 Glyphosate is a broad-spectrum, nonselective systemic herbicide. It kills a plant by inhibiting a specific enzyme pathway and interferes with the synthesis of aromatic amino acids necessary for its growth.

3 For a critical and comprehensive policy analysis of USAID-led alternative development policies in Colombia, see Vargas Meza (2010). USAID also funded small infrastructural projects, such as building bridges, school installations, and landing strips, extending rural electricity, and making road improvements. Coca-growing communities have also been considered "vulnerable populations" for which the state should provide social assistance programs, such as Familias en Acción, Familias Guardabosques, Empleo en Acción, and Jóvenes en Acción.

4 With roots in the 1980s, the AUC grew to an estimated 31,000 militants. The umbrella organization formed in April 1997, and was heavily financed through the drug trade, landowners, cattle ranchers, mining and oil companies, multinational corporations, and Colombia's traditional political class. For more on the relationship between government forces and paramilitaries in Putumayo, the role of the FARC-EP with respect to the profitability of drug trafficking, and local results of the 2006 demobilization of the AUC, see Jansson (2008).

5 The larger project, San Miguel Mira hacia Colombia y el Mundo, was developed over four years between 2004 and 2008. It was coordinated and financed by CINEP in alliance with the municipal government of San Miguel, another Jesuit NGO, Fundación Social, and the social investment money of ECOPETROL, the largest and primary petroleum company in Colombia. Besides the farm school, the project included a youth communications collective called Young Reporters of San Miguel, the development of a "Sub-Regional Public Agenda for Bajo Putumayo," and a political campaign called Vote por la Amazonia (Vote for the Amazon) (CINEP 2007).

6 For an intimate portrayal of the daily life of coca growers and workers, see my creative ethnographic nonfiction and photographic installation, *Fresh Leaves* (Lyons 2014a).

7 Margarita Chaves (1998) elucidates the everyday interactions, permeating identities, productive frictions, and political alliances between indigenous and nonindigenous communities in Putumayo despite classic representations of their supposed isolation and conflicting forms of conviviality.

8 Constitutional reforms enacted since 2004, including intellectual property laws that the Colombian government was required to adopt when signing the free trade agreement with the United States (Law 1518 of April 2012), have steadily declared a variety of campesino and indigenous food production, commercialization, and seed-propagating practices to be illegal. For example, Resolution 000957 of April 2008 prohibited the production, breeding, and commercialization of creole hens; Resolution 002546 of 2004 prohibited the production and commercialization of artisanal brown sugar (*panela*); Decree 1500 of 2007 prohibited the production and butchering of livestock in municipal capitals and small towns, privatized the sale of refrigerated meat, and granted agro-industrial companies the sole right to sell raw milk (Decree 2838 of 2006). For more information on these neoliberal reforms and their discontents, see www.semillas.org.co.

9 The Environmental Clinic is an initiative founded and supported by the Quito-

based NGO Acción Ecológica, a leading environmental organization in Latin America that was founded twenty-five years ago.

10 For a comprehensive history of the Mesa Regional, see Mesa Regional de Organizaciones Sociales (2015).

11 MEROS activists have come to reject what Winifred Tate (2013) calls "proxy citizenship" by distancing themselves from NGOs that interfere with their ability to make direct claims for redress from the state.

12 For more information on the 2013 National Agrarian, Ethnic, and Popular Strike and post-strike political processes and negotiations, see Mesa Nacional Agropecuaria y Popular de Interlocucción y Acuerdo (MIA) (2015).

13 For details about the rural participatory methodology, focus, and themes of the formulation of the Andean-Amazonian Integral Development Plan (PLADIA 2035), see MEROS (2017).

14 Heraldo's first appointment as secretary of agriculture, which lasted for six months, occurred due to the influence of the leftist Alianza Democrática M-19 (M-19 Democratic Alliance) political party in Putumayo. In 1989, a majority of M-19 (19th of April) guerrillas demobilized and transitioned into the legal political sphere to become the M-19 party. The party united with other political movements, including the demobilized Popular Liberation Army (EPL), Revolutionary Workers' Party, and the Quintín Lamé, to become the April 19 Democratic Alliance Movement in 1991. According to the Comisión Andina de Juristas (1993), the AD M-19 became a space of regional encounter between leftist community and union leaders. Given the presence of the Masetos paramilitaries in Putumayo, the AD M-19 also became an alternative to the Union Patriótica (UP) political movement that was linked to the FARC and suffered genocidal persecution by paramilitaries throughout the country. His second appointment as secretary in January 2016 occurred when Sorrel Aroca Rodríguez was elected Putumayo's first female governor. Sorrel was elected for the Green Party, which represents a convergence of center, center-left, and moderate conservative sectors in Putumayo. During her campaign, Sorrel built political alliances with the Regional Working Group of Social Organizations (MEROS), which is largely composed of campesino organizations and unions.

15 According to the most recent national census, conducted in 2005, 76 percent of Putumayo's population is categorized as mestizo, 18 percent indigenous, and 6 percent black or Afro-Colombian.

16 The global militarization of the drug war in the 1980s became more explicitly intertwined after 9/11 when US foreign aid to Colombia conflated counternarcotic and counterterrorism wars (see Ramírez 2005).

17 Herding time is described as a common and shared time, different from the previous flow of time that is established by and also creates the flock (Despret and Meuret 2016).

18 Raffles agrees with the critical formulation of feminist geographer Doreen Massey, who writes about places as meeting points of social relations, as the outcomes of difference and inequality, and also as producing difference and

inequality (Massey, Allen, and Cochrane 1998). Places are never stable, but are constantly being made through the ongoing affective, imaginative, discursive relationships and physical labor between humans and nonhumans.

19 I am grateful to my colleague Tania Pérez-Bustos for connecting me back to Freire's critical pedagogy.

20 This triad is situated in a genealogy of Latin American critical theory that informs my work, which includes dependency theory, liberation theology, participatory action research, and a current of thought or movement that is sometimes referred to as *pensamiento latinoamericano en ciencia, tecnología y sociedad* (Latin American thinking on science, technology and society).

21 See also Kirksey (2015) for work on "emergent ecologies" and finding possibilities in the wreckage of ongoing disasters through the symbiotic associations of plants, animals, and microbes that flourish in what he conceives of as unexpected places.

CHAPTER 2: THE THEATER OF LIFE IS ALSO A STAGE OF DEATH

1 Povinelli makes an argument that biopolitics always rests on the governance of the division between life and nonlife, or what she calls *geontopower*, rather than simply in and on the governance of and through life. When Povinelli refers to the division between life and nonlife, she is not "primarily referring to that which had life and now does not, but a more foundational division within Western governance between that which supposedly arrives into existence inert and that which arrives with an active potentiality" (Povinelli, Coleman, and Yusoff 2017, 5). I thank my colleague Pierre Du Plessis for pointing me to this interview.

2 In Colombia's post-accord and transitional justice process, environmental coalitions and social movements, such as CENSAT Agua Viva and Ríos Vivos, have proposed reconstructing the environmental or biocultural memory of territories that have been epicenters of war and the extractive and mega-infrastructural projects that the social and armed conflict created the conditions to implement through displacement and dispossession of local communities. See http://www.contagioradio.com/memoria-ambiental-el-desafio-de-enunciar-la-vida-articulo-47833/, accessed on March 17, 2018.

3 In the historical accounts of the development of Colombian soil science, biology and microbiology did not begin to grow in importance until the 1980s (Orozco and Medina 2005). For a more generalized historical account of the relegation of soil science to lower status in histories of microbiology, which tended to emphasize the latter's contribution to molecular biology, see Strick (2014).

4 The rise of soil science was closely correlated with the demand for increased soil fertility to support capitalist agriculture in Europe and North America; the demand for "cheap" food that both provided the caloric fuel and resulted from successive agricultural revolutions (Moore 2010); guano imperialism (Clark and Bellamy Foster 2009); the introduction of synthetic fertilizers, and in the post–World War II era, the global expansion of the Green Revolution's emphasis on "improved" seeds and agrichemical input substitution.

5 See Jenny (1941).

6 All of these statistics were taken from Julián Serna Giralda's public presentation, "2009 Year of Soils in Colombia—Campaign for Protection and Recuperation," given on May 16, 2009.

7 See "What if the World's Soils Runs Out?" *Time,* http://world.time.com/2012/12 /14/what-if-the-worlds-soil-runs-out/, accessed on February 14, 2017.

8 Colombia's Ministry of the Environment, Housing, and Territorial Development declared June 17th National Soil Day in Resolution 0170.

9 See also Logan (1995).

10 Soil microbiologists inform us that one gram of soil may harbor up to 10 billion microorganisms of possibly thousands of different species, far exceeding the microbial diversity of eukaryotic organisms. Given that less than 1 percent of microorganisms observed under a microscope are cultivated and characterized, soil ecosystems are, to a large extent, unknown (Torsvik and Øvreås 2002). See the September 2008 issue of *National Geographic,* "Where Food Begins," the May 2010 issue of *Nature Geoscience,* "Microbial Mitigation of Soil Carbon Emission," and the 2004 special issue of *Science* on "Soils—The Final Frontier."

11 According to the legal analysis of Mesa Cuadros, Sánchez Supelano, and Silva Porras, soil has been an object of regulation according to multiple sectorial or partial visions. For example, norms have been issued for soils as: (1) a territorial component; (2) a substrate of agricultural development (agricultural soil Articles 178–80 CNNR), mining (Law 685 of 2001), housing (Law 388 of 1997 and CONPES 3583), and infrastructure; or (3) property that should be conserved and protected (2015, 104).

12 Many thanks to Javier Vanegas for the insightful conversations we shared during my time at IBUN. This particular quote was taken from a conversation we had on May 14, 2010.

13 For a comprehensive overview of applications of bioaugmentation, biostimulation, and biocontrol with plant growth–promoting organisms and organic material amendments for the improvement of soil biology and fertility, see Singh et al. (2011).

14 I thank Radhika Subramanian and Rane Willerslev for their provocative questions and comments during a one-day workshop at the Parsons School for Constructed Environments and GIDEST at the New School, where I first presented a draft of this chapter.

15 Tania Pérez-Bustos, María Fernanda Olarte Sierra, and Adriana Díaz del Castillo (2014) focus on the case of women geneticists working in the field of forensic genetics in Colombia to make an opposite analysis about the feminization of the scientific disciplines of microbiology and bacteriology. I thank Carolina Olivera for this statistic on the SCSS.

16 I also take inspiration from Natasha Myers's (2015b) ethnographic exploration of how biologists who model proteins develop bodily intuitions about the movement of molecules, crafting a habitus that has their fingers, hands, and bodies

responding to, and miming, computer models of the protein structures they study.

17 Soils are studied in twelve different agronomy departments in the country, and as an elective in chemistry, biology, geology, forestry engineering, and agricultural engineering degrees. The study of soils is also conducted by the research centers associated with the coffee, rice, cacao, cane sugar, banana, flower, and African palm oil industries (Burbano and Silva Mojica 2010, x).

18 This quote was taken from a conversation I had with Botero on August 11, 2010.

19 See "El 65,8% de la tierra apta para sembrar en Colombia no se aprovecha," *El Tiempo*, http://m.eltiempo.com/economia/sectores/el-658-de-la-tierra-apta-para-sembrar-en-colombia-no-se-aprovecha/16601436, accessed on February 22, 2017.

20 See https://www.oxfamamerica.org/static/media/files/rr-divide-and-purchase-land-concentration-colombia-211013-en.pdf, accessed on May 24, 2018.

21 Optical mineralogy is the study of geological materials by measuring their optical properties in order to help reveal their origin and evolution. Most commonly, soil samples are prepared as thin sections for study in the laboratory with a petrographic microscope. The principal difference between a petrographic and a biological microscope is the presence of two polarizing elements, one above and one below the stage.

22 I thank my friend and colleague Emilie Dionne for suggesting Diprose's work to me.

23 See Letey et al. (2003) for a critical assessment of the conceptual weaknesses and contradictions in the concept of "soil quality." See Laishram et al. (2012) for a review essay on the development of soil health approaches as distinct from soil quality information.

24 See Engel-Di Mauro (2014) for a discussion about the importance of conceptualizing soil quality and degradation in ways that account for social relations of domination alongside biophysical and wider ecological processes.

25 I was provoked to think more about the specific effects of bombs and shelling on soils when I came across Joe Hupy's concept of "soil bombturbation," which he developed to explore a specific anthropogenic mode of soil displacement and mixing produced by the impacts of aerial or in situ explosive devices. Besides being connected to militarized intervention, what he calls soil bombturbation may also be related to invasive and extractive industrial activities (Hupy and Koehler 2012; Hupy and Schaetzl 2006).

26 See, for example, http://www.futurefarmers.com/#projects/soilkitchen. See also Landa and Feller (2010).

27 This quote is taken from a personal email communication with Cortés on April 25, 2011.

28 See, for example, Cortés (1990, 2004).

29 See Zeiderman (2016) for an in-depth ethnography of the self-built settlements of Bogotá's urban periphery, and how state actors and citizens navigate environmental hazards, including soils categorized as geologically unstable.

30 According to the Presidential Program for the Integral Action against Land-

mines (AICMA), 10,253 victims were reported between 1990 and February 2013. Thirty-eight percent of cases involved civilians and 62 percent members of the armed forces. See the official webpage of the government's Directorate for the Integral Action against Landmines for updated statistics, legislation, and humanitarian demining initiatives in Colombia: http://www.accioncontraminas .gov.co/Paginas/AICMA.aspx.

CHAPTER 3: PARTIAL ALLIANCES AMONG MINOR PRACTICES

1 Raffles and WinklerPrins refer to the work of late US anthropologist Julian Steward and Smithsonian archaeologist Betty Meggers between the 1940s and 1960s in establishing the Amazon as the principal ethnographic proving ground for cultural ecological theories of environmental determination. For more on this racialized debate, see Kawa (2016).

2 See, for example, Glaser and Woods (2004); Mann (2006); Morcote-Ríos (2008).

3 In the settler colonial context of the United States, Traci Brynne Voyles (2015) analyzes the discursive emptying of Navajo rangeland and the violent making of racialized and gendered bodies and degraded landscapes that ensued.

4 For more details on PRORADAM and contemporary experiences of indigeneity in the western Colombian Amazon in relation to state multiculturalism and environmentalism, see Del Cairo (2012).

5 See, for example, the Congressional Research Service report for Congress, "Drug Crop Eradication and Alternative Development in the Andes," published on November 18, 2005, https://www.everycrsreport.com/reports/RL33163.html, accessed on March 28, 2017.

6 In effect, anyone engaged in agriculture, livestock raising, forestry, or infrastructure projects and activities that can affect the soils is legally obliged to carry out conservation, recuperation, and/or compensation practices that are determined according to regional characteristics. These criteria were reiterated in Law 99 of 1993, which entrusted the now Ministry of Environment and Sustainable Development with issuing a "statute for the suitable use zonification of territory for its appropriate ordinance and the national regulations for the use of the soil concerning its environmental aspects." Law 388 of 1997 establishes that all territorial ordinances should be founded on the following principles: the social and ecological function of property, the prevalence of public interest over private interest, and the equitable distribution of burdens and benefits (Mesa Cuadros, Sánchez Supelano, and Silva Porras 2015, 106).

7 Law 30, passed in Colombia in 1986, criminalized the cultivation of marijuana, coca, and opium poppies in excess of twenty plants. This placed small growers, who account for around 70 percent of coca cultivation, and large-scale traffickers in the same legal category by ignoring the structural forces that lead people to settle in rural frontier areas and engage in illicit livelihoods in the first place. Furthermore, the Colombian government permitted the aerial fumigation of its national parks and large swaths of its biodiverse tropical forests.

8　See https://www.semana.com/contenidos-editoriales/hidrocarburos-son-el
-futuro/articulo/putumayo-clave-para-el-futuro-petrolero-del-pais/590016
?fbclid=iwar1mlahui1n5vhlzo4bxm3jj3d_rywlsqpmpqstyaz33kgizxu1yewpdmeg,
accessed on May 26, 2019.

9　See https://www.elespectador.com/noticias/economia/paz-nos-va-permitir
-sacar-mas-petroleo-de-zonas-vedadas-articulo-627058, accessed on July 14,
2018.

10　The former Colombian Institute of Rural Development (INCODER) identified
four territorial conflicts hindering the state's institutional capacities to adju-
dicate land titles in Putumayo: (1) conflicts where various actors are disputing
control of the same territory; (2) difficulties in accessing the formal procedures
to legalize land ownership; (3) confusion over borders and limits between existing territories; and (4) infrastructural and/or extractive projects that prevent
local communities from obtaining property titles (Centro Nacional de Memoria
Histórica 2015, 49).

11　See http://www.elespectador.com/noticias/judicial/18-batallones-protegeran
-infraestructura-energetica-articulo-366470, accessed on February 10, 2017.

12　Writing on the ways Afro-descendants have progressively lost control of their
lives and territories, activist and anthropologist Carlos Rosero says, "If war is the
continuation of the economy by other means," it is clear in Colombia that "inde-
pendently of who wields the weapons, they are used to enforce societal and de-
velopmental logics that are completely at odds with those of the ethnic groups"
(2002, 550).

13　See http://noticias.caracoltv.com/colombia/llegaron-primeros-cien-guerrilleros
-zona-veredal-en-putumayo, accessed on January 15, 2017, and http://www.bbc
.com/news/world-latin-america-38826392, accessed on February 2, 2017.

14　See http://colombia2020.elespectador.com/pais/cartas-desde-la-marcha-final,
accessed on February 5, 2017.

15　According to a news report by Global Witness, of the eighty-seven human
rights defenders murdered in Latin America in 2016, sixty were defending rights
linked to environmental destruction. Disturbingly, these statistics likely under-
represent the scale of the problem, as many killings of defenders and activists
around the world go unreported. See http://www.aida-americas.org/blog/as
-killings-increase-how-can-we-defend-the-defenders, accessed on August 7,
2017.

16　This is the number of pre-accords signed as of July 2017. See http://miputumayo
.com.co/2017/07/27/campesinos-del-putumayo-logran-acuerdos-de-sustitucion
-de-cultivos/, accessed on July 27, 2017. Also, http://www.elespectador.com
/noticias/nacional/putumayo-da-un-paso-mas-para-empezar-sustitucion
-de-cultivos-ilicitos-articulo-705257, accessed on July 29, 2017.

17　See http://pacifista.co/preguntenle-a-las-mujeres-cocaleras-ellas-tienen-las
-respuestas/, accessed on May 12, 2017.

18　See, for example, http://www.elespectador.com/noticias/judicial/militares

-policias-y-fiscalia-se-unen-perseguir-los-des-articulo-674320, accessed on January 28, 2017.

19 See https://sostenibilidad.semana.com/medio-ambiente/articulo/deforestacion-en-colombia-despues-del-acuerdo-de-paz-con-las-farc/41088, accessed on July 14, 2018.

20 See, for example, Wilches-Chaux (2012) and United Nations Colombia and German Cooperation (2014).

21 See, for example, Kregg Hetherington's (2013) ethnographic response to the statement that soy kills ("la soja mata"), a refrain often repeated by campesino activists living on the edge of Paraguay's rapidly expanding soybean frontier. See also the special issue of the *Journal of Political Ecology*, "Production/ Destruction in Latin America," edited by Javiera Barandiaran and Casey Walsh (2017).

22 Like many areas of rural Colombia, the Media Bota Caucana experiences complex and unresolved land titling controversies that involve Colombia's National Parks Office, the National Land Agency (previously known as INCODER), oil and industrial mining companies and concessions, indigenous communities, and small farmers. See Jeremy Campbell's (2015) ethnographic treatment of the way colonists have been drawn to the Brazilian Amazon and engage in different speculative practices to "conjure" property and make land claims.

23 Nitrogen, phosphorus, and potassium are considered the most important macronutrients required for plant, and 10-30-10 is a widely sold commercial concentration of N-P-K.

24 Here I also think with my feminist science studies colleague Tania Pérez-Bustos (2017a), who describes *calado* embroidery not as the subservience of the embroiderer to the fabrics and threads, but, along with Leanne Prain, as "an active and concentrated state in which relationalities between humans and non-human actors are interwoven" (f).

25 Heraldo Vallejo completed an MA degree in territorial planning and environmental management in November 2016; his thesis focused specifically on the "Influence of the Application of Organic Material in the Recovery of Degraded Soils in the Amazonian Region." See Vallejo (2016).

26 Latour wants to know how the sciences can at the same time be realist and constructivist, immediate and intermediary. In this particular chapter of *Pandora's Hope*, he asks what the spoken word refers to when scientists speak of soil, and how what he calls a "circulating reference" is produced through constant substitutions of the world that scientists both make and encounter.

27 For a historical case study of the tensions surrounding the implementation of US soil taxonomy in the Colombian coffee sector, see Tally (2006).

28 Barad writes that "to be entangled is not simply to be intertwined, as in the joining of separate and preexisting entities, but to lack an independent, self-contained existence outside of the relation itself" (2007, ix).

29 See also Forsyth's (2011) example about the contentious application of the USDA's universal soil loss equation (USLE) in the highlands of northern Thailand.

30 See http://www.foodnewslatam.com/6654-análisis-de-suelo, -la-mano-derecha -de-los-agricultores-colombianos.html, accessed on February 28, 2017. I thank my colleague Julio Arias Vanegas for drawing my attention to this article.

CHAPTER 4: DECOMPOSITION AS LIFE POLITICS

1 For details on Resolution 970, see http://www.ica.gov.co/Normatividad/Normas -Ica/Resoluciones-Oficinas-Nacionales/RESOLUCIONES-DEROGADAS /RESOL-970-DE-2010.aspx, accessed on May 6, 2017.

2 A chiva is a rustic artisan bus that has been adapted to mountainous geography, and that continues to be used as rural public transport in Colombia.

3 To think historically about the subversive potential of gardens, see Jill Casid's compelling eighteenth- and nineteenth-century imperial plantation histories of what she calls "countercolonial gardens" and "black resistance landscapes" in the Caribbean (2005, 191). See also Bettina Stoetzer's (2018) work on garden practices and alternative "ruderal ecologies" that are both cultivated and "wildly" emerge from Berlin's ruins.

4 The most well documented case of this kind of criminalization occurred in 2013 after the Colombian Agricultural and Livestock Institute (ICA), accompanied by the Mobile Anti-Riot Squad (ESMAD), confiscated and destroyed sixty-two to seventy tons of rice being transported by rice growers in Campoalegre, Huila. ICA officials justified the seizure based on two arguments: (1) that the majority of the rice was uncertified, and hence deemed a possible phytosanitary risk, and (2) that the rice was not being transported in its original packaging, but instead in recycled flour and fertilizer sacks that were considered to be potential contaminants. See also https://justiciaambientalcolombia.org/2017/05/04/semillas -de-colombia-en-peligro-documental-9-70-de-victoria-solano/, accessed on May 5, 2017.

5 See Myers (2015a) for provocative reflections on the ways plants sense and make sense of their worlds.

6 See Mol, Moser, and Polis (2010) for more on care as an experimental everyday practice and the matter of "tinkering" on farms.

7 Angela Garcia (2017) speaks to this in her articulation of people's daily efforts to maintain and increase their hold on life, while acknowledging the uncertainty and risk of such efforts. Her ethnography explores the way "death deeply penetrates life, not only in the sense of diminishing it, but also in the sense of giving it new resources to survive, perhaps, even flourish" (318) among heroin-addicted kin.

8 I am indebted to Alberto Corsín Jiménez's (2013) discussion of the urban cultivation table produced in handmade urbanism workshops in Madrid as a means of imagining how a city, or in this case a territory, may jump ahead of itself.

9 Hugh Raffles writes that "the power of the jungle is that it can never be tamed: once tamed, it's no longer jungle" (2004, 235).

10 I refer to James Scott's (2009) discussion about crop choice as "escape agriculture." He argues that particular crops, such as roots and tubers, along with specific agricultural techniques, such as shifting cultivation, exhibit characteristics that make them more or less resistant to raiding, state-making, and state appropriation. He claims that these crops and techniques have historically been employed as an "agropolitical strategy" (193) among disparate upland Southeast Asian peoples.

11 Rancière (2012) elaborates on similar questions in his historical reflection on the labor struggles and desires of nineteenth-century workers. He opens up concepts of labor and workers' movements beyond the idea of rebellion against specific hardships and conditions to also consider longings for the possibility to live other lives.

CHAPTER 5: RESONATING FARMS AND VITAL SPACES

1 For more information on the zrc, see the Cartilla Pedagógica de las Zonas Reserva Campesina en Colombia, https://issuu.com/anzorc/docs/cartillazonas dereservacampesina, and newspaper articles by Alfredo Molano, for example, http://www.elespectador.com/opinion/opinion/ciudadanos-campesinos -columna-669782, and http://www.semana.com/nacion/articulo/el-lio-zonas -reserva-campesina/337007-3.

2 For an excellent analysis of the need for agrarian jurisdiction, and a case study about territorial conflicts in the department of Cauca, see http://lasillavacia .com/silla-llena/red-rural/historia/los-conflictos-territoriales-y-la-urgencia -de-una-jurisdiccion, accessed on August 4, 2017.

3 See Besky and Padwe (2016) for ways to think with plants about territory.

4 I am compelled by Stenger's insistence that to hope does not "mean hope for one thing or another thing or as a calculated attitude, but to try and feel and put into words a possibility for becoming" (2002, 245).

5 Henry Salgado Ruiz (2012) makes a similar argument that entrepreneurial or agro-industrial territorial visions and the territorial conceptions of campesinos overlapped, but never canceled one another out in the Colombian Amazon. Grubačić and O'Hearn conceptualize these enactments in terms of "exilic spaces" (2016, 1) or place-based modes of exit populated by communities striving to relatively escape from both state regulation and capitalist accumulation.

6 See, for example, how the Zapatistas explain their practices of autonomy as tangible, fluid, processual, and self-organized. See http://enlacezapatista.ezln .org.mx for "Gobierno autónomo I," "Gobierno autónomo II," "Resistencia y autónoma," and "Participación de las mujeres en el gobierno autónomo."

7 See also socivil (2010).

8 Resonating with my reference to concepts and practices of *buen vivir* (living well), Viveiros de Castros writes, "There is no better than enough" (2013, 37).

9 The Revolutionary Youth of Colombia was a youth organization affiliated with the Worker's Party of Colombia (ptc). The Worker's Party has its origins in the

Marxist-Maoist-Leninist Worker's Independent and Revolutionary Movement (MOIR), founded in 1969, and from which the PTC eventually separated in 1999. For more historical information about the PTC, see the official party website, http://partidodeltrabajodecolombia.org.

10 The concept of selvacino/a draws upon the historically situated spatial, material, and symbolic dimensions of diverse campesino struggles in Colombia. See exemplary studies by Fajardo Montaña (2002); LeGrand (1988); Palacios (2011); and Tovar-Pinzón (1995) on the rise of a campesino identity in Colombia during the nineteenth and twentieth centuries that is deeply tied to territorial struggles.

11 Todd (2016) makes an important intervention on the ways indigenous thinkers/ practitioners have been at the forefront of conceptualizing this relationality while only sometimes or minimally being cited by posthumanist, ontoepistemic, and new feminist materialist scholars.

12 Classic technical classifications of water systems in the Amazon are based on three categories: (1) white waters (rivers that originate in the Andes and that are "naturally" rich in minerals due to the characteristics of their soils of origin); (2) black waters (rivers that originate in the Amazon basin itself, and that are considered less "naturally" fertile; rather than acquiring nutrients from the minerals stored in soils, they depend on falling organic matter—seeds, leaves, and insects, for example); and (3) crystal waters (those rivers or streams that originate at high altitudes, where there is not necessarily a notable amount of organic material in the soils).

13 I base this reflection on personal conversations I had with the geologist Mauricio Valencia Zepulveda and the agrologist Pedro Botero, who have long-term experience living and working in the department of Putumayo.

14 Mol analytically explores "what it is to hang together" through the empirical material of clafoutis. She poses lessons learned about coherent composites, which maintain tension, heterogeneity, precarity, and robustness.

15 Inspired by Viveiros de Castro's perspectivism, Kohn describes the way the Runa of Ecuador's upper Amazon are aware of the selfhood of the many beings that people the cosmos in which they live. He writes: "They hold that their ability to enter this web of relations—to be aware and to relate to other selves [for example, creatures that they hunt]—depends on the fact that they share this quality with the other beings that make up this ecology" (2013, 17).

16 See Quesada Tovar et al. (2015) for a legal analysis and case study about the particular conditions of campesino communities in regard to prior consultation norms in Colombia.

17 This pressure escalated in 2008 when Colombian President Álvaro Uribe bombed a FARC-EP camp on Ecuadorian soil during Operación Fénix, killing FARC leader Raúl Reyes, twenty other FARC members, and four Mexican students. The attacked was guided by radars located on the US military base in Manta, Ecuador. See www.telesurtv.net/news/7-anos-del-ataque-uribista-a-Ecuador-20150301-0047.html, accessed on July 30, 2017.

1   I thank generative conversations with colleagues at the Volatile Futures/Earthly Matters conference held at Bennington College in May 2017 for inspiring these thoughts on soils in relation to planetary scales, especially the comments of Andrea Ballestero.

2   I take inspiration from what Claire Colebrook (2017) calls a thought experiment or scenario of the "Anthropocene counterfactual" to imagine a world where humans came into being but did not develop technology to the point where the geological impact of the Anthropocene took place.

3   I thank my colleague Rima Praspaliauskiene for encouraging me to write this ethnographic vignette on hands.

REFERENCES

Alaimo, Stacy, and Susan Hekman, eds. 2008. *Material Feminisms*. Indianapolis: Indiana University Press.

Área de Memoria Histórica, Línea de Investigación Tierra y Conflicto (AMH). 2009. *El despojo de tierras y territorios. Aproximación conceptual.* Bogotá: AMH, Comisión Nacional de Reparación y Reconciliación, IEPRI.

Ariza, Eduardo, María Clemencia Ramírez, and Leonardo Vega. 1998. *Atlas Cultural de la Amazonia Colombiana: La Construcción del Territorio en el Siglo XX.* Bogotá: Instituto Colombiano de Antropología.

Asociación Ambiente y Sociedad. 2019. *Petróleo en la Amazonia: ¿Pueblos Indígenas en Peligro?* Ambiente y Sociedad: Bogotá.

Bachelard, Gaston. 2002. *Earth and Reveries of Will: An Essay on the Imagination of Matter.* Dallas: Dallas Institute.

Barad, Karen. 2003. "Posthumanist Performativity: Toward an Understanding of How Matter Comes to Matter." *Signs: Journal of Women in Culture and Society* 28, no. 3: 801–31.

Barad, Karen. 2007. *Meeting the Universe Halfway.* Durham, NC: Duke University Press.

Barad, Karen. 2010. "Quantum Entanglements and Hauntological Relations of Inheritance: Dis/Continuities, SpaceTime Enfoldings, and Justice-to-Come." *Derrida Today* 3, no. 2: 240–68.

Barandiaran, Javiera, and Casey Walsh, eds. 2017. "Production/Destruction in Latin America," Special issue of the *Journal of Political Ecology* 24.

Baronnet, Bruno, Mariana Mora Bayo, and Richard Stahler-Sholk, eds. 2011. *Luchas "muy otras": Zapatismo y autonomía en las comunidades indígenas de Chiapas.* Mexico City: Universidad Autónoma Metropolitana.

Bataille, Georges. 1993. *The Accursed Share: An Essay on General Economy*, vol. 2: *The History of Eroticism.* New York: Zone.

Baveye, Philippe, Astrid R. Jacobson, Suzanne E. Allaire, John P. Tandarich, and Ray B. Bryant, 2006. "Whither Goes Soil Science in the United States and Canada?" *Soil Science* 171: 501–18.

Beittel, June S. 2012. "Colombia: Background, U.S. Relations, and Congressional Interest." Congressional Research Service, RL3225.

Besky, Sarah, and Jonathon Padwe. 2016. "Placing Plants in Territory." *Environment and Society: Advances in Research* 7: 9–28.

Biehl, João. 2005. *Vita: Life in a Zone of Social Abandonment*. Berkeley: University of California Press.

Blaser, Mario. 2009. "The Threat of the Yrmo: The Political Ontology of a Sustainable Hunting Program." *American Anthropologist* 111, no. 1: 10–20.

Bonilla, Daniel. 1969. *Sirvos de Dios y Amos de Indios: El Estado y la Misión Capuchina en el Putumayo*. Bogotá: Author.

Bowker, Geoffrey, and Susan Leigh Star. 2000. *Sorting Things Out: Classification and Its Consequences*. Cambridge, MA: MIT Press.

Burbano, Hernán. 2010. "El suelo y su importancia para la sociedad." In *Ciencia del suelo: Principios Básicos*, edited by Hernán Burbano and Francisco Silva Mojica, 553–94. Bogotá: Sociedad Colombiana de la Ciencia del Suelo.

Burbano, Hernán, and Francisco Silva Mojica, eds. 2010. *Ciencia del suelo: Principios básicos*. Bogotá: Sociedad Colombiana de la Ciencia del Suelo.

Butler, Judith, Zeynep Gambetti, and Leticia Sabsay. eds. 2016. *Vulnerability in Resistance*. Durham, NC: Duke University Press.

Calle, M. C. 2014. "Putumayo está en crisis." *Semana*, August 30.

Campbell, Jeremy. 2015. *Conjuring Property: Speculation and Environmental Futures in the Brazilian Amazon*. Seattle: University of Washington Press.

Carroll, Leah Anne. 2011. *Violent Democratization: Social Movements, Elites, and Politics in Colombia's Rural War Zones, 1984–2008*. Notre Dame, IN: University of Notre Dame Press.

Casid, Jill. 2005. *Sowing Empire: Landscape and Colonization*. Minneapolis: University of Minnesota Press.

Castro-Gómez, Santiago. 2005. *La poscolonialidad explicada a los niños*. Popayán, Colombia: Instituto Pensar, Universidad Javeriana, Jigra de letras—Editorial Universidad del Cauca.

Centro Nacional de Memoria Histórica. 2015. *Petróleo, Coca, Despojo Territorial, y Organización Social en Putumayo*. Bogotá: CNMH.

Centro Nacional de Memoria Histórica. 2016. *Tierras y Conflictos Rurales: Historia, Políticas Agrarias y Protagonistas*. Bogotá: CNMH.

Chaves, Margarita. 1998. "Identidad y representación entre indígenas y colonos de la Amazonia occidental." In *Modernidad, Identidad y Desarrollo*, edited by María Lucía Sotomayor, 273–86. Bogotá: ICANH-Colciencias.

Choy, Timothy. 2011. *Ecologies of Comparison: An Ethnography of Endangerment in Hong Kong*. Durham, NC: Duke University Press.

Churchman, G. Jock. 2010. "The Philosophical Status of Soil Science." *Geoderma* 157, no. 3–4: 214–21.

CINEP. 2007. *Escuela de Desarrollo Amazónico Sostenible*. Bogotá: Ediciones Antropos.

Clark, Brett, and John Bellamy Foster. 2009. "Ecological Imperialism and the Global Metabolic Rift: Unequal Exchange and the Guano/Nitrates Trade." *International Journal of Comparative Sociology* 50, no. 3–4: 311–34.

Clark, Nigel. 2011. *Inhuman Nature: Sociable Life on a Dynamic Planet*. London: SAGE.

Clastres, Pierre. 2007. *Society against the State*. New York: Zone.

Clastres, Pierre. 2010. *Archeology of Violence*. Los Angeles: Semiotext.

Clínica Ambiental. 2009. "Huerta para la Soberanía Alimentaria en la Región Amazónica." *Alerta Naranja* 3 (April).

Cohen, Benjamin. 2009. *Notes from the Ground: Science, Soil and Society in the American Countryside*. New Haven, CT: Yale University Press.

Colebrook, Claire. 2017. "We Have Always Been Post-Anthropocene: The Anthropocene Counterfactual." In *Anthropocene Feminism*, edited by Richard Grusin, 1–20. Minneapolis: University of Minnesota Press.

Comisión Andina de Juristas. 1993. *Putumayo*. Serie Informes Regionales de Derechos Humanos. Bogotá: Códice Editorial Ltda.

Connolly, William. 2010. "Materialities of Experience." In *New Materialisms: Ontology, Agency and Politics*, edited by Diana Coole and Samantha Frost, 178–200. Durham, NC: Duke University Press.

CORPOAMAZONIA. 2007. *Atlas Ambiental del Putumayo*. Bogotá: Amaranta Ltda.

Corsín Jiménez, Alberto. 2013. "Three Traps Many." Paper presented at the John E. Sawyer Seminar "Indigenous Cosmopolitics: Dialogues about the Reconstitution of Worlds," University of California, Davis, March 18.

Cortés, Abdón. 1990. "El suelo y la biodiversidad." *Revista La Tadeo* 27.

Cortés, Abdón. 1991a. "El suelo, gran ausente en nuestra agenda ambiental." *Revista La Tadeo* 30.

Cortés, Abdón. 1991b. "El suelo, maravilloso teatro de la vida." *Revista La Tadeo* 31.

Cortés, Abdón. 2004. *Suelos Colombianos: Una Mirada Desde la Academia*. Bogotá: Fundación Universidad de Bogotá Jorge Tadeo Lozano.

Cortés, Abdón, and Celso Ibarra. 1981. *Los Suelos de la Amazonia Colombiana: Criterios para la utilización racional*. Bogotá: IGAC.

Craib, Raymond. 2004. *Cartographic Mexico: A History of State Fixations and Fugitive Landscapes*. Durham, NC: Duke University Press.

Cruickshank, Julie. 2005. *Do Glaciers Listen? Local Knowledges, Colonial Encounters, and Social Imagination*. Seattle: University of Washington Press.

Culma, Edinso. 2013. "Militares, parentesco y la construcción del estado local en Leguízamo (Putumayo)." MA thesis, Facultad de Ciencias Sociales, Quito.

D'Alisa, Giacomo, Federico DeMaria, and Giorgos Kallis. 2015. *Degrowth: A Vocabulary for a New Era*. London: Routledge.

de la Cadena, Marisol. 2010. "Indigenous Cosmopolitics in the Andes: Conceptual Reflections beyond 'Politics.'" *Cultural Anthropology* 25, no. 2: 334–70.

de la Cadena, Marisol. 2015a. *Earth Beings: Ecologies of Practice across Andean Worlds*. Durham, NC: Duke University Press.

de la Cadena, Marisol. 2015b. "Uncommoning Nature." *E-Flux Journal*, 56th Biennale. http://supercommunity.e-flux.com/authors/marisol-de-la-cadena/.

de la Cadena, Marisol, and Marianne E. Lien, eds. 2015. "Anthropology and STS: Generative Interfaces, Multiple Locations." HAU: *Journal of Ethnographic Theory* 5, no. 1: 437–75.

Del Cairo, Carlos. 2012. "Environmentalizing Indigeneity: A Comparative Ethnography on Multiculturalism, Ethnic Hierarchies and Political Ecology in the Colombian Amazon." PhD dissertation, University of Arizona, Tucson.

Del Cairo, Carlos, Iván Montenegro-Perini, and Juan Sebastián Vélez. 2014. "Naturalezas, subjetividades y políticas ambientales en el Noroccidente amazónico: Reflexiones metodológicas para el análisis de conflictos socioambientales." *Boletín de Antropología* 29, no. 48: 13–40.

Deleuze, Gilles, and Félix Guattari. 1987. *A Thousand Plateaus: Capitalism and Schizophrenia*. London: Continuum.

Deleuze, Gilles, and Félix Guattari. 1996. *What Is Philosophy?* New York: Columbia University Press.

Delgado, Ana, and Israel Rodríguez-Giralt. 2014. "Creole Interferences: A Conflict over Biodiversity and Ownership in the South of Brazil." In *Beyond Imported Magic: Essays on Science, Technology, and Society in Latin America*, edited by Eden Medina, Ivan da Costa Marques, and Christina Holmes, 331–48. Cambridge, MA: MIT Press.

Departamento Nacional de Planeación de Colombia. 2010. "Plan nacional de desarrollo, 2010–2014: Prosperidad para todos." Bogotá. https://www.dnp.gov.co/Plan-Nacional-de-Desarrollo/PND-2010–2014/Paginas/Plan-Nacional-De-2010–2014.aspx.

DeSilvey, Caitlin. 2006. "Observed Decay: Telling Stories with Mutable Things." *Journal of Material Culture* 11, no. 3: 318–38.

DeSilvey, Caitlin. 2017. *Curated Decay: Heritage Beyond Saving*. Minneapolis: University of Minnesota Press.

Despret, Vinciane, and Michel Meuret. 2016. "Cosmoecological Sheep and the Arts of Living on a Damaged Planet." *Environmental Humanities* 8, no. 1: 24–36.

Diprose, Rosalyn. 2002. *Corporeal Generosity: On Giving with Nietzsche, Merleau-Ponty, and Levinas*. New York: State University of New York Press.

Duarte, Carlos. 2016. *Desencuentros territoriales: Caracterización de los conflictos en las regiones de la altillanura, Putumayo y Montes de María*. Bogotá: Instituto Colombiano de Antropología e Historia.

Duarte, Carlos. 2017. "Los conflictos territoriales y la urgencia de una Jurisdicción Agraria." *LaSillaVacia*. http://lasillavacia.com/silla-llena/red-rural/historia/los-conflictos-territoriales-y-la-urgencia-de-una-jurisdiccion.

Dumit, Joseph. 2012. "Prescription Maximization and the Accumulation of Surplus Health in the Pharmaceutical Industry: The_BioMarx_Experiment." In *Lively Capital: Biotechnologies, Ethics, and Governance in Global Markets*, edited by Kaushik Sunder Rajan, 45–92. Durham, NC: Duke University Press.

Engel-Di Mauro, Salvatore. 2014. *Ecology, Soils, and the Left: An Ecosocial Approach.* New York: Palgrave Macmillan.

Escobar, Arturo. 1994. *Encountering Development: The Making and Unmaking of the Third World.* Princeton, NJ: Princeton University Press.

Escobar, Arturo. 2007. "Worlds and Knowledges Otherwise: The Latin American Modernity/Coloniality Research Program." *Cultural Studies* 21, no. 2–3: 179–210.

Escobar, Arturo. 2008. *Territories of Difference: Place, Movements, Life, Redes.* Durham, NC: Duke University Press.

Escobar, Arturo. 2014. *Sentipensar con la tierra: Nuevas lecturas sobre desarrollo, territorio y diferencia.* Medellín, Colombia: UNALUA.

Fajardo Montaña, Dario. 2002. *Para Sembrar la Paz Hay que Aflojar la Tierra: Comunidades, tierras y territorios en la construcción de un país.* Bogotá: Universidad Nacional de Colombia.

Fearnside, Philip M. 1985. "Environmental Change and Deforestation in the Brazilian Amazon." In *Change in the Amazon Basin: Man's Impact on Forests and Rivers,* edited by J. Hemming, 70–89. Manchester, UK: Manchester University Press.

Federici, Silvia. 2018. *Re-Enchanting the World: Feminism and the Politics of the Commons.* Cambridge, MA: MIT Press.

Forsyth, Tim. 2011. "Politicizing Environmental Explanations: What Can Political Ecology Learn from Sociology and Philosophy of Science?" In *Knowing Nature: Conversations at the Intersection of Political Ecology and Science Studies,* edited by Mara Goldman, Paul Nadasy, and Matthew Turner, 31–46. Chicago: University of Chicago Press.

Fortun, Kim. 2001. *Advocacy after Bhopal: Environmentalism, Disaster, New Global Orders.* Chicago: University of Chicago Press.

Franco, Fernando. 2006. "La Corporación Araracuara y La Colonización Científica de las Selvas Ecuatoriales Colombianas." *Revista Colombia Amázonica,* 13–34.

Freire, Paulo. 1970. *Pedagogy of the Oppressed.* New York: Herder and Herder.

Garcia, Angela. 2017. "Death as a Resource for Life." In *Living and Dying in the Contemporary World: A Compendium,* edited by Veena Das and Clara Han, 316–28. Oakland: University of California Press.

García Márquez, Gabriel. 1972. *Leaf Storm.* New York: Harper and Row.

Gibson-Graham, J. K. 2006a. *The End of Capitalism (As We Knew It).* Minneapolis: University of Minnesota Press.

Gibson-Graham, J. K. 2006b. *Postcapitalist Politics.* Minneapolis: University of Minnesota Press.

Giraldo, Isis. 2016. "Coloniality at Work: Decolonial Critique and the Postfeminist Regime." *Feminist Theory* 17, no. 2: 157–73.

Glaser, Bruno, and William Woods, eds. 2004. *Amazonian Dark Earths: Explorations in Space and Time.* Berlin: Springer.

Goldman, Mara, Paul Nadasy, and Mathew Turner, eds. 2011. *Knowing Nature: Conversations at the Intersection of Political Ecology and Science Studies.* Chicago: University of Chicago Press.

González, Roberto. 2001. *Zapotec Science: Farming and Food in the Northern Sierra of Oxaca*. Austin: University of Texas Press.

Graeter, Stefanie. 2017. "To Revive an Abundant Life: Catholic Science, Neoextractivist Politics in Peru's Mantaro Valley." *Cultural Anthropology* 32, no. 1: 117–48.

Green, Lesley, ed. 2013. *Contested Ecologies: Dialogues in the South on Nature and Knowledge*. Cape Town: Human Sciences Research Council.

Grubačić, Andrej, and Denis O'Hearn. 2016. *Living at the Edges of Capitalism: Adventures in Exile and Mutual Aid*. Oakland: University of California Press.

Gudynas, Eduardo. 2014. "Sustentación, aceptación y legitimación de los extractivismos: Múltiples expresiones pero un mismo basamento." OPERA 14: 137–59.

Gudynas, Eduardo, and Alberto Acosta. 2011. "El buen vivir más allá que el desarrollo." *Quehacer* 181: 70–81.

Gutiérrez, Raúl. 2006. *A desordenar! Por una historia abierta de la lucha social*. Mexico City: Juan Pablos-CEAM.

Gutiérrez Escobar, Laura. 2017. "Seed Sovereignty Struggles in an Emberá-Chami Community in Colombia." *Alternautas: (Re)Searching Development: The Abya Yala Chapter*. Accessed July 14, 2018. http://www.alternautas.net/blog/2017/9/4/seed-sovereignty-struggles-in-an-ember-cham-community-in-colombia.

Güttler, Nils. 2015. "Drawing the Line: Mapping Cultivated Plants and Seeing Nature in Nineteenth Century Plant Geography." In *New Perspectives on the History of Life Sciences and Agriculture*, edited by Denise Phillips and Sharon Kingsland, 27–52. New York: Springer.

Hacking, Ian. 1983. *Representing and Intervening: Introductory Topics in the Philosophy of Natural Science*. Cambridge: Cambridge University Press.

Hage, Ghassan. 2012. "Critical Anthropological Thoughts and the Radical Political Imaginary Today." *Critique of Anthropology* 32, no. 3: 258–308.

Haraway, Donna J. 2008. *When Species Meet*. Minneapolis: University of Minnesota Press.

Haraway, Donna J. 2016. *Staying with the Trouble: Making Kin in the Chthulucene*. Durham, NC: Duke University Press.

Haraway, Donna J., and Martha Kenney. 2015. "Anthropocene, Capitalocene, Chthulucene." In *Art in the Anthropocene: Encounters among Aesthetics, Politics, Environments and Epistemologies*, edited by Heather Davis and Etienne Turpin, 255–70. London: Open Humanities.

Harding, Susan. 2008. *Sciences from Below: Feminisms, Postcolonialities, and Modernities*. Durham, NC: Duke University Press.

Harrison, Paul. 2008. "Corporeal Remains: Vulnerability, Proximity and Living On after the End of the World." *Environment and Planning A* 40, no. 2: 423–45.

Hartemink, Alfred, and Alex McBratney. 2010. "A Soil Science Renaissance." *Geoderma* 148, no. 2: 123–29.

Hartigan, John. 2017. *Care of the Species: Races of Corn and the Science of Plant Biodiversity*. Minneapolis: University of Minnesota Press.

Harvey, Penelope, and Hannah Knox. 2015. *Roads: An Anthropology of Infrastructure and Expertise*. Ithaca, NY: Cornell University Press.

Hayden, Cori. 2003. *When Nature Goes Public: The Making and Unmaking of Bioprospecting in Mexico*. Princeton, NJ: Princeton University Press.

Heller, Chaia. 2007. "Techne versus Technoscience: Divergent (and Ambiguous) Notions of Food 'Quality' in the French Debate over GM Crops." *American Anthropologist* 109, no. 4: 603–15.

Helmreich, Stefan. 2009. *Alien Ocean: Anthropological Voyages in Microbial Seas*. Berkeley: University of California Press.

Hetherington, Kregg. 2013. "Beans before the Law: Knowledge Practices, Responsibility, and the Paraguayan Soy Boom." *Cultural Anthropology* 28, no. 1: 65–85.

Hole, Francis D. 1988. "The Pleasures of Soil Watching." *Orion Nature Quarterly* (spring): 6–11.

Hupy, Joseph, and Thomas Koehler. 2011. "Modern Warfare as a Significant Zoogeomorphic Disturbance upon the Landscape." *Geomorphology* 157–58: 169–82.

Hupy, Joseph, and Randall Schaetzl. 2006. "Introducing 'Bombturbation,' a Singular Type of Soil Disturbance and Mixing." *Soil Science* 171, no. 11: 823–36.

Hustak, Carla, and Natasha Myers. 2012. "Involutionary Momentum: Affective Ecologies and the Sciences of Plant/Insect Encounters." *Differences: A Journal of Feminist Cultural Studies* 23, no. 3: 74–118.

Ingold, Tim. 2011. *Being Alive: Essays on Movement, Knowledge and Description*. London: Routledge.

Ingold, Tim, and Jo Lee Vergunst, eds. 2008. *Ways of Walking: Ethnography and Practice on Foot*. Hampshire, UK: Ashgate.

Instituto Geográfico Agustín Codazzi (IGAC). 2008. *Suelos para Niños*. Bogotá: IGAC.

James, William. 1996. *A Pluralistic Universe*. Lincoln: University of Nebraska Press.

Jansson, Oscar. 2008. "The Cursed Leaf: An Anthropology of the Political Economy of Cocaine Production in Southern Colombia." PhD dissertation, Uppsala University.

Jaquette Ray, Sarah, Jay Sibara, and Stacey Alaimo, eds. 2017. *Disability Studies and the Environmental Humanities: Toward an Eco-Crip Theory*. Lincoln: University of Nebraska Press.

Jenny, Hans. 1941. *Factors of Soil Formation: A System of Quantitative Pedology*. New York: McGraw-Hill.

Jenny, Hans. 1980. *The Soil Resource: Origin and Behavior*. New York: Springer-Verlag.

Jenny, Hans. 1984. "My Friend the Soil: A Conversation with Hans Jenny." *Journal of Soil and Water Conservation* (May–June): 158–61.

Kawa, Nicholas. 2016. *Amazonia in the Anthropocene: People, Soils, Plants, Forests*. Austin: University of Texas Press.

Kirksey, Eben. 2015. *Emergent Ecologies*. Durham, NC: Duke University Press.

Klein, Naomi. 2014. *This Changes Everything: Capitalism vs. Climate Change*. New York: Simon and Schuster.

Kohn, Eduardo. 2013. *How Forests Think: Toward an Anthropology beyond the Human*. Berkeley: University of California Press.

Laishram, Joylata, K. G. Saxena, Rakesh K. Maikhuri, and Kottapalli S. Rao. 2012. "Soil Quality and Soil Health: A Review." *International Journal of Ecology and Environmental Sciences* 38, no. 1: 19–37.

Landa, Edward, and Christian Feller, eds. 2010. *Soil and Culture*. New York: Springer.

Latour, Bruno. 1988. *The Pasteurization of France*. Cambridge, MA: Harvard University Press.

Latour, Bruno. 1999. *Pandora's Hope: Essays on the Reality of Science Studies*. Cambridge, MA: Harvard University Press.

Lavelle, Phillip. 2000. "Ecological Challenges for Soil Science." *Soil Science* 165, no. 1: 73–86.

Law, John. 2004. *After Method: Mess in Social Science Research*. London: Routledge.

Law, John, and Wen-yuan Lin. 2015. "Provincialising STS: Postcoloniality, Symmetry and Method." 2015 Bernal Prize Plenary at the Annual Meeting of the Society for the Social Studies of Science, Denver, CO, November 11–15.

LeGrand, Catherine. 1988. *Colonización y protesta campesina en Colombia 1850–1950*. Bogotá: Universidad Nacional de Colombia.

León, Tomás. 1999. "Perspectivas de la investigación en los suelos de la Amazonia." In *Amazonia Colombiana: Diversidad y Conflicto*, edited by Germán Andrade, Adriana Hurtado, and Ricardo Torres, 237–55. Bogotá: AGORA.

Letey, John, Robert E. Sojka, Dana R. Upchurch, D. Keith Cassel, Kenneth R. Olson, William A. Payne, Steven E. Petrie, Graham H. Price, Robert J. Reginato, H. Don Scott, Philip J. Smethurst, and Glover B. Triplett. 2003. "Deficiencies in the Soil Quality Concept and Its Application." *Journal of Soil and Water Conservation* 58, no. 4: 180–87.

Lévi-Strauss, Claude. 1966. *The Savage Mind*. Chicago: University of Chicago Press.

Li, Fabiana. 2015. *Unearthing Conflict: Corporate Mining, Activism, and Expertise in Peru*. Durham, NC: Duke University Press.

Logan, William Bryant. 1995. *Dirt: The Ecstatic Skin of the Earth*. New York: W. W. Norton.

Lyons, Kristina. 2014a. *Fresh Leaves*. Creative Ethnographic Nonfiction and Photographic Installation, published by the Centre for Imaginative Ethnography's Galleria, York University, May 14. http://imaginativeethnography.org/galleria/fresh-leaves-by-kristina-lyons/.

Lyons, Kristina. 2014b. "Soil Science, Development, and the 'Elusive Nature' of Colombia's Amazonian Plains." *Journal of Latin American and Caribbean Anthropology* 19, no. 2: 212–36.

Lyons, Kristina. 2016a. "Decomposition as Life Politics: Soils, *Selva*, and Small Farmers under the Gun of the U.S.-Colombian War on Drugs." *Cultural Anthropology* 31, no. 1: 55–80.

Lyons, Kristina. 2016b. "Selva Life and Death: A Conversation in Images with Kristina Lyons." *Cultural Anthropology*. March 4. https://culanth.org/fieldsights/814-selva-life-and-death-a-conversation-in-images-with-kristina-lyons.

Lyons, Kristina. 2018. "Chemical Warfare in Colombia, Evidentiary Ecologies, and *Senti-actuando* [Feeling Acting] Practices of Justice." *Social Studies of Science* 48, no. 3: 414–37.

Malagón, Dimas. 2005. "Pasado, Presente y Futuro de la Ciencia del Suelo en Colombia: Análisis de Aspectos Fundamentales." *Suelos Ecuatoriales* 35, no. 1: 78–102.

Mann, Charles. 2006. *1491: New Revelations of the Americas before Columbus*. New York: Vintage.

Marx, Karl. 1990. *Capital*, vol. 1. London: Penguin Classics.

Marx, Karl. 1991. *Capital*, vol. 3. London: Penguin Classics.

Massey, Doreen. 2005. *For Space*. London: SAGE.

Massey, Doreen, John Allen, and Allan Cochrane. 1998. *Rethinking the Region*. London: Routledge.

Mathews, Andrew. 2011. *Instituting Nature: Authority, Expertise, and Power in Mexican Forests*. Cambridge, MA: MIT Press.

Mbembe, Achille. 2017. *Critique of Black Reason*. Durham, NC: Duke University Press.

McLean, Stuart. 2009. "Stories and Cosmogonies: Imagining Creativity beyond 'Nature' and 'Culture.'" *Cultural Anthropology* 24, no. 2: 213–45.

Medina, Eden, Iván da Costa Marques, and Christina Holmes, eds. 2014. *Beyond Imported Magic: Essays on Science, Technology, and Society in Latin America*. Cambridge, MA: MIT Press.

Mesa Cuadros, Gregorio, Luis Fernando Sánchez Supelano, and Yazmin Andrea Silva Porrras. 2015. "Aportes Conceptuales para la Construcción de una Política de Gestión Ambiental de Suelos (GEAS) en Colombia." In *Conflictividad ambiental y afectaciones a derechos ambientales*, edited by Gregorio Mesa Cuadros, 95–123. Bogotá: Universidad Nacional de Colombia.

Mesa Nacional Agropecuaria y Popular de Interlocución y Acuerdo (MIA). 2015. *Del Paro Nacional Agraria a la Cumbre Agraria, Campesina, Etnica y Popular 2013–2014: Sistematización del proceso*. Bogotá: MIA.

Mesa Regional de Organizaciones Sociales del Putumayo, Baja Bota Caucana y Cofanía, Jardines de Sucumbíos (Nariño). 2015. *Putumayo: Sembrando vida y construyendo identidad. Historia de la Mesa Regional (2006–2014)*. Bogotá: Corcas Editores SAS.

Mesa Regional de Organizaciones Sociales del Putumayo, Baja Bota Caucana y Cofanía, Jardines de Sucumbíos (Nariño). 2017. PLADIA 2035. *Plan de Desarrollo Integral AndinoAmazónico 2035*. Mexico City: S&D Soporte y Diseño.

MINGA. 2008. *Informe Misión de Observación a la Situación de Derechos Humanos en el Bajo Putumayo*. Bogotá: Ediciones Versailles.

Mol, Annemarie. 2002. *The Body Multiple: Ontology in Medical Practice*. Durham, NC: Duke University Press.

Mol, Annemarie. 2016. "Clafoutis as a Composite: On Hanging Together Felicitously." In *Modes of Knowing: Resources from the Baroque*, edited by John Law and E. Rupert, 242–65. Manchester: Mattering Press.

Mol, Annemarie, Ingunn Moser, Jeannette Polis, eds. 2010. *Care in Practice: On Tinkering in Clinics, Homes and Farms.* New Brunswick, NJ: Transaction.

Montgomery, Nick, and Carla Bergman. 2017. *Joyful Militancy: Building Thriving Resistance in Toxic Times.* Chico, CA: AK Press.

Moore, Jason. 2010. "The End of the Road: Agricultural Revolutions in the Capitalist World—Ecology 1450–2010." *Journal of Agrarian Change* 10, no. 3: 389–413.

Morcote-Ríos, Gasper. 2008. *Antiguas Habitantes en Ríos de Aguas Negras: Ecosistemas y Cultivos y el Interfluvio Amazonas-Putumayo Colombia-Brasil.* Bogotá: Universidad Nacional de Colombia.

Murphy, Michelle. 2006. *Sick Building Syndrome and the Problem of Uncertainty: Environmental Politics, Technoscience, and Women Workers.* Durham: Duke University Press.

Myers, Natasha. 2015a. "Conversations on Plant Sensing: Notes from the Field." *Nature Culture* 3: 35–66.

Myers, Natasha. 2015b. *Rendering Life Molecular: Models, Modelers, and Excitable Matter.* Durham, NC: Duke University Press.

Myers, Natasha. 2017. "From the Anthropocene to the Planthroposcene: Designing Gardens for Plant/People Involution." *History and Anthropology* 28, no. 3: 297–301.

Myers, Natasha. 2019. "From Edenic Apocalypse to Gardens against Eden: Plants and People in and after the Anthropocene." In *Infrastructure, Environment, and Life in the Anthropocene*, edited by Kregg Hetherington. Durham, NC: Duke University Press.

Nadasy, Paul. 2003. *Hunters and Bureaucrats: Power, Knowledge and Aboriginal-State Relations in the Southwest Yukon.* Vancouver: University of British Columbia Press.

Nader, Laura. 1972. "Up the Anthropologist: Perspectives Gained from Studying Up." In *Reinventing Anthropology*, edited by Dell Hymes, 284–311. New York: Pantheon.

Nouvet, Elysée. 2014. "Some Carry On, Some Stay in Bed: (In)Convenient Affects and Agency in Neoliberal Nicaragua." *Cultural Anthropology* 29, no. 1: 80–102.

Ojeda, Diana, and María Camila González. 2018. "Elusive Space: Peasants and Resource Politics in the Colombian Caribbean." In *Land Rights, Biodiversity Conservation and Justice*, edited by Sharlene Mollet and Thembela Kepe, 89–106. New York: Routledge.

Orozco, Francisco Hernando, and Marisol Medina. 2005. "Pasado y Futuro de la Microbiología del Suelos en Colombia." *Suelos Ecuatoriales* 35, no. 1: 124–36.

Palacio, Germán. 2004. *Civilizando la tierra caliente: la supervivencia de los bosquesinos amazónicos 1850–1930.* Bogotá: ASCUN.

Palacios, Marco. 2011. *¿De quién es la tierra? Propiedad, politización y protesta campesina en la década de 1930.* Bogotá: Universidad de los Andes.

Paley, Dawn. 2014. *Drug War Capitalism.* Oakland, CA: AK Press.

Papadopoulos, Dimitris. 2010. "Activist Materialism." *Deleuze Studies* 4: 64–83.

Papadopoulos, Dimitris. 2014. "Politics of Matter: Justice and Organization in Technoscience." *Social Epistemology* 28, no. 1: 70–85.

Papadopoulos, Dimitris, Niamh Stephenson, and Vassilis Tsianos. 2008. *Escape Routes: Control and Subversion in the 21st Century*. London: Pluto.

Pérez-Bustos, Tania. 2017a. "Thinking with Care: Unraveling and Mending in an Ethnography of Craft Embroidery and Technology." *Revue d´anthropologie des connaissances* 11, no. 1: a–u.

Pérez-Bustos, Tania. 2017b. "A Word of Caution toward Homogenous Appropriations of Decolonial Thinking in STS." *Catalyst: Feminism, Theory, Technoscience* 3, no. 1: 39–41.

Pérez-Bustos, Tania, María Fernanda, Olarte Sierra, and H. Adriana Díaz del Castillo. 2014. "Working with Care: Narratives of Invisible Women Scientists Practicing Forensic Genetics in Colombia." In *Beyond Imported Magic: Essays on Science, Technology, and Society in Latin America*, edited by Eden Medina, Ivan da Costa Marques, and Christina Holmes, 67–84. Cambridge, MA: MIT Press.

Petryna, Adriana. 2002. *Life Exposed: Biological Citizenship after Chernobyl*. Princeton, N.J.: Princeton University Press.

Pollan, Michael. 2006. *The Omnivore's Dilemma: A Natural History of Four Meals*. New York: Penguin.

Povinelli, Elizabeth. 1995. "Do Rocks Listen? The Cultural Politics of Apprehending Australian Aboriginal Labor." *American Anthropologist* 97, no. 3: 505–18.

Povinelli, Elizabeth. 2011a. *Economies of Abandonment: Social Belonging and Endurance in Late Liberalism*. Durham, NC: Duke University Press.

Povinelli, Elizabeth. 2011b. "The Woman on the Other Side of the Wall: Archiving the Otherwise in Postcolonial Digital Archives." *Journal of Feminist Cultural Studies* 22, no. 1: 146–71.

Povinelli, Elizabeth. 2016. *Geontologies: A Requiem to Late Liberalism*. Durham, NC: Duke University Press.

Povinelli, Elizabeth, Mathew Coleman, and Kathryn Yusoff. 2017. "An Interview with Elizabeth Povinelli: Geontopower, Biopolitics and the Anthropocene." *Theory, Culture and Society* 34, no. 2–3: 169–85.

Primavesi, Ana. 1984. *Manejo Ecológico del Suelo: La agricultura en regiones tropicales*. Buenos Aires: Editorial El Ateneo.

PRORADAM. 1979. *La Amazonia colombiana y sus recursos*. Bogotá: Proyecto Radargramétrico del Amazonas—PRORADAM.

Puig de la Bellacasa, María. 2014. "Encountering Bioinfrastructure: Ecological Struggles and the Sciences of Soil." *Social Epistemology: A Journal of Knowledge, Culture and Policy* 28, no. 1: 26–40.

Puig de la Bellacasa, María. 2015a. "Ecological Thinking, Material Spirituality, and the Poetics of Infrastructure." In *Boundary Objects and Beyond: Working with Leigh Star*, edited by Geoffrey C. Bowker, Stefan Timmermans, Ellen Balka, and Adele E. Clarke, 47–68. Cambridge, MA: MIT Press.

Puig de la Bellacasa, María. 2015b. "Making Time for Soil: Technoscientific Futurity and the Place of Care." *Social Studies of Science* 45, no. 5: 691–716.

Quesada Tovar, Carlos, Carlos Olaya Díaz, Luis Fernando Sánchez Supelano, and José Agustín Labrador Forero. 2015. "Consulta Previa a Comunidades Campesinas Como Sujetos Culturales Diferenciados." In *Derechos ambientales en disputa: Algunos estudios de caso sobre conflictividad ambiental*, edited by Gregorio Mesa Cuadros, 211–40. Bogotá: Universidad Nacional de Colombia.

Raffles, Hugh. 2002. *In Amazonia: A Natural History*. Princeton, NJ: Princeton University Press.

Raffles, Hugh. 2004. "Jungle." In *Patterned Ground: Entanglements of Nature and Culture*, edited by Stephen Harrison, Steve Pile, and Nigel Thrift, 234–35. London: Reaktion.

Raffles, Hugh, and Antoinette WinklerPrins. 2003. "Further Reflections on Amazonian Environmental History: Transformation of Rivers and Streams." *Latin American Research Review* 38, no. 3: 165–87.

Rajão, Raoni, Ricardo Duque, and Rahul De. 2014. "Introduction: Voices from within and outside the South: Defying sts Epistemologies, Boundaries, and Theories." *Science, Technology and Human Values* 39, no. 6: 1–6.

Ramírez, María Clemencia. 1996. *Frontera Fluida Entre Andes, Piedemonte y Selva: El Caso del Valle de Sibundoy, Siglos XVI–XVII*. Bogotá: Editorial abc Limitada.

Ramírez, María Clemencia. 2001. *Entre el estado y la guerrilla: Identidad y ciudadanía en el movimiento de los campesinos cocaleros del Putumayo*. Bogotá: icanh.

Ramírez, María Clemencia. 2005. "Aerial Spraying and Alternative Development in Plan Colombia: Two Sides of the Same Coin or Two Contested Policies?" *ReVista: Harvard Review of Latin America* (spring/summer): 54–57.

Ramírez, María Clemencia, Ingrid Bolívas, Juliana Iglesias, María Clara Torres, and Teófilo Vásquez. 2010. *Elecciones, coca, conflicto y partidos políticos en Putumayo 1980–2007*. Bogotá: cinep Programa por la Paz, icanh.vv.

Rancière, Jacques. 2012. *Proletarian Nights: The Worker's Dream in Nineteenth-Century France*. London: Verso.

Ritz, Karl. 2014. "Life in Earth: A Truly Epic Production." In *The Soil Underfoot: Infinite Possibilities for a Finite Resource*, edited by G. Jock Churchman and Edward R. Landa, 379–94. New York: crc Press.

Rose, Nikolas. 2007. *The Politics of Life Itself: Biomedicine, Power, and Subjectivity in the Twenty-First Century*. Princeton, N.J.: Princeton University Press.

Rosero, Carlos. 2002. "Los afrodescendientes y el conflicto armado en Colombia: La insistencia en lo propio como alternativa." In *Afrodescendientes en las Américas: Trayectorias sociales e identitarias*, edited by Claudia Mosquera, Mauricio Pardo, and Odile Hoffmann, 547–60. Bogotá: Universidad Nacional/icanh.

Roy, Brototi, and Joan Martínez Alier. 2017. "Blokadia por la justicia clímatica." *Ecología Política* 53 Antropoceno. http://www.ecologiapolitica.info/? p=9770.

ruv. 2016. "Registro Único de Víctimas." Unidad para las Víctimas, República de Colombia. Bogotá. http://rni.unidadvictimas.gov.co/RUV.

Sagan, Dorion. 2011. "The Human Is More than Human: Interspecies Communities and the New 'Facts of Life.'" Society for Cultural Anthropology, November 18. https://culanth.org/fieldsights/the-human-is-more-than-human-interspecies -communities-and-the-new-facts-of-life.

Salgado Ruiz, Henry. 2012. "El campesinado de la Amazonia colombiana: Construcción territorial, colonización forzada y resistencias." PhD dissertation, University of Montreal, Montreal, Quebec.

Santoyo, Álvaro Andrés. 2002. "Representaciones nacionales de la Amazonia colombiana, 1900–1975. Una aproximación antropológica e histórica a la retórica y la política de la producción de la subjetividad y la naturaleza." Final research report presented to the Ministry of Culture, Bogotá. Manuscript.

Schmink, Marianne, and Charles H. Wood. 1992. *Contested Frontiers in Amazonia.* New York: Columbia University Press.

Scott, James. 2009. *The Art of Not Being Governed: An Anarchist History of Upland Southeast Asia.* New Haven, CT: Yale University Press.

Semillas de Paz: La Obra de El Padre Alcides Jiménez en el Putumayo. 1996. Puerto Caicedo, Putumayo.

Serje, Margarita. 2011. *El Revés de la Nación: Territorios Salvajes, Fronteras y Tierras de Nadie.* Bogotá: Universidad de los Andes.

Serres, Michel. 1995a. *Genesis.* Ann Arbor: University of Michigan Press.

Serres, Michel. 1995b. *The Natural Contract.* Ann Arbor: University of Michigan Press.

Serres, Michel. 2007. *The Parasite.* Minneapolis: University of Minnesota Press.

Shepherd, Chris. 2005. "Imperial Science: The Rockefeller Foundation and Agricultural Science in Peru, 1940–1960." *Science as Culture* 14, no. 2: 113–37.

Shiva, Vandana. 1993. *Monocultures of the Mind: Perspectives on Biodiversity and Biotechnology.* London: Zed.

Simpson, Leanne. 2011. *Dancing on Our Turtle's Back: Stories of Nishnaabeg Re-Creations, Resurgence, and a New Emergence.* Winnipeg: ARP Books.

Singh, Ajay, Nagina Parmar, Ramesh Kuhad, and Owen Ward. 2011. "Bioaugmentation, Biostimulation, and Biocontrol in Soil Biology." In *Bioaugmentation, Biostimulation, and Biocontrol,* edited by Ajay Singh, Nagina Parmar, and Ramesh Kuhad, 1–23. London: Springer.

SOCIVIL. 2010. "Memorias del Foro Colombo-Ecuatoriano: La Cuchara. Por la Autonomía, la Soberanía y la Seguridad Alimentaria." Mocoa, Putumayo: Author.

Star, Susan Leigh. 1995. *Ecologies of Knowledge: Work and Politics in Science and Technology.* Albany: State University of New York Press.

Star, Susan Leigh. 2006. "Whose Infrastructure Is It, Anyway?" Unpublished paper.

Stengers, Isabelle. 2002. *Penser avec Whitehead: Une libre et sauvage création de concepts.* Paris: Le Seuil.

Stengers, Isabelle. 2005a. "The Cosmopolitical Proposal." In *Making Things Public: Atmospheres of Democracy,* edited by Bruno Latour and Peter Weibel, 994–1003. Cambridge, MA: MIT Press.

Stengers, Isabelle. 2005b. "Including Nonhumans in Political Theory: Opening Pandora's Box." In *Political Matter: Technoscience, Democracy, and Public Life*, edited by Bruce Braun and Sarah Whatmore, 3–33. Minneapolis: University of Minnesota Press.

Stengers, Isabelle. 2005c. "Introductory Notes on an Ecology of Practices." *Cultural Studies Review* 11, no. 1: 183–96.

Stengers, Isabelle. 2017. "Autonomy and the Intrusion of Gaia." *South Atlantic Quarterly* 116, no. 2: 381–400.

Stengers, Isabelle, and Philippe Pignarre. 2011. *Capitalist Sorcery: Breaking the Spell.* New York: Palgrave MacMillan.

Stevenson, Lisa. 2014. *Life beside Itself: Imagining Care in the Canadian Artic.* Berkeley: University of California Press.

Stewart, Katie. 2007. *Ordinary Affects.* Durham, NC: Duke University Press.

Stoetzer, Bettina. 2018. "Ruderal Ecologies: Rethinking Nature, Migration, and the Urban Landscape in Berlin." *Cultural Anthropology* 33, no. 2: 295–323.

Stoler, Ann. 2016. *Duress: Imperial Durabilities in Our Times.* Durham, NC: Duke University Press.

Strick, James. 2014. "The Cycle of Life Concept, Soil Microbiology and Soil Science Restored to the History of Ecology." *Studies in History and Philosophy of Biological and Biological Sciences* 48: 119–21.

Sunder Rajan, Kaushik. 2006. *Biocapital: The Constitution of Postgenomic Life.* Durham, NC: Duke University Press.

Svampa, Maristella. 2012. "Consenso de los commodities, giro ecoterritorial y pensamiento crítico en América Latina." *Observatorio Social de América Latina* 32: 15–38.

TallBear, Kim. 2014. "Standing with and Speaking as Faith: A Feminist-Indigenous Approach to Inquiry." *Journal of Research Practice* 10, no. 2. http://jrp.icaap.org/index.php/jrp/article/view/405/371.

Tally, Rebecca. 2006. "A Young and Dynamic Country: Soil Science and Conservation in Colombia, circa 1950." Paper presented at the New York State Latin American History Workshop, October 15.

Tate, Winifred. 2013. "Proxy Citizenship and Transnational Advocacy: Colombian Activists from Putumayo to Washington, DC." *American Ethnologist* 40, no. 1: 55–70.

Tate, Winifred. 2015. *Drugs, Thugs, and Diplomats: U.S. Policymaking in Colombia.* Stanford, CA: Stanford University Press.

Taussig, Michael. 1984. "Culture of Terror, Space of Death: Roger Casement's Putumayo Report and the Explanation of Torture." *Comparative Studies in Society and History* 26, no. 3: 467–97.

Taussig, Michael. 1987. *Shamanism, Colonialism and the Wild Man: A Study in Terror and Healing.* Chicago: University of Chicago Press.

Tilley, Helen. 2011. *Empire, Development, and the Problem of Scientific Knowledge, 1870–1950: Africa as a Living Laboratory.* Chicago: University of Chicago Press.

Todd, Zoe. 2016. "Indigenizing the Anthropocene." In *Art in the Anthropocene: Encounters among Aesthetics, Politics, Environment and Epistemologies*, edited by Heather Davis and Etienne Turpin, 241–54. London: Open Humanities Press.

Torsvik, Vigdis, and Lise Øvreås. 2002. "Microbial Diversity and Function in Soil: From Genes to Ecosystems." *Current Opinion in Microbiology* 5: 240–45.

Tovar-Pinzón, Hernes. 1995. *Que nos tengan en cuenta. Colonos, empresarios y aldeas: Colombia 1800–1900*. Bogotá: Colcultura-Tercer Mundo Editores.

Tsing, Anna. 2011. "Arts of Inclusion: How to Love a Mushroom." *Australian Humanities Review* 50: 19.

Tsing, Anna. 2015. *The Mushroom at the End of the World: On the Possibility of Life in Capitalist Ruin*. Princeton, NJ: Princeton University Press.

Tuck, Eve, and Marcia McKenzie. 2015. *Place in Research: Theory, Methodology, and Methods*. New York: Routledge.

Ulloa, Astrid. 2016. "Feminismos territoriales en América Latina: Defensas de la vida frente a los extractivismos," NÓMADAS 45: 123–39.

United Nations Colombia and German Cooperation. 2014. *Consideraciones ambientales para la construcción de una paz territorial estable, duradera y sostenible en Colombia: Insumos para la discusión*. ReliefWeb. https://reliefweb.int/report /colombia/consideraciones-ambientales-para-la-construccion-de-una-paz -territorial-estable.

United Nations Office on Drugs and Crime (UNODC). 2005. "Colombia: Coca Cultivation Survey for 2004." https://www.unodc.org/ unodc/en/crop-monitoring /?tag=Colombia.

USDA-NRCS. 2010. "Soil Quality/Soil Health Concepts." http://soils.usda.gov/sqi /assessment/assessment.html.

Vallejo, Heraldo. 1993a. *El Murciélago Humano*. Villagarzón, Putumayo: Author.

Vallejo, Heraldo. 1993b. "El nuevo hombre amazónico: Una visión del desarrollo para el departamento del Putumayo." *Putumayo: Expresión de Identidad Regional* 1: 8–25.

Vallejo, Heraldo. 2016. "Influencia de la Aplicación de Materia Orgánica en la Recuperación de Suelos Degradados en la Región Amazónica," MA thesis, Universitat de Barcelona and Centro Universitario Internacional de Barcelona, Spain.

Vallejo, Heraldo, Alberto Campaña, and Jairo Muchavisoy. 2002. *Capacitación y Asistencia Técnica Cultivos Agrícolas*. Villagarzón, Putumayo: Author.

Vargas Meza, Ricardo. 1999. *Fumigación y conflicto: Políticas antidrogas y deslegitimación del estado en Colombia*. Bogotá: TNI/Acción Andina.

Vargas Meza, Ricardo. 2010. *Desarrollo alternativo en Colombia y participación social: Propuestas hacia un cambio de estrategia*. Bogotá: Corcas Editores Ltda.

Veltmeyer, Henry, and James Petras. 2014. *The New Extractivism: A Post-Neoliberal Development Model or Imperialism of the Twenty-First Century?* London: Zed.

Verran, Helen. 2001. *Science and an African logic*. Chicago: University of Chicago Press.

Verran, Helen. 2002. "A Postcolonial Moment in Science Studies: Alternative Firing

Regimes of Environmental Scientists and Aboriginal Landowners." *Social Studies of Science* 32, no. 5–6: 729–62.

Verran, Helen. 2013. "Engagements between Disparate Knowledge Traditions: Toward Doing Difference Generatively and in Good Faith." In *Contested Ecologies: Dialogues in the South on Nature and Knowledge*, edited by Lesley Green, 141–61. Cape Town: Human Sciences Research Council.

Villa, William, and Juan Houghton. 2005. *Violencia política contra los pueblos indígenas, 1974–2004*. Bogotá: Grupo Internacional de Trabajo sobre Asuntos Indígenas, Organización Indígena de Antioquia (OIA), Centro de Cooperación al Indígena.

Viveiros de Castro, Eduardo. 2004. "Perspectival Anthropology and the Method of Controlled Equivocation." *Tipití* 2, no. 1: 3–22.

Viveiros de Castro, Eduardo. 2013. "Economic Development and Cosmopolitical Re-Involvement: From Necessity to Sufficiency." In *Contested Ecologies: Dialogues in the South on Nature and Knowledge*, edited by Lesley Green, 28–41. Cape Town: Human Sciences Research Council.

Vora, Kalindi. 2015. *Life Support: Biocapital and the New History of Outsourced Labor*. Minneapolis: University of Minnesota Press.

Voyles, Traci Brynne. 2015. *Wastelanding: Legacies of Uranium Mining in Navajo Country*. Minneapolis: University of Minnesota Press.

Wilches-Chaux, Gustavo. 2012. "Supongamos la paz con la naturaleza." *El Tiempo*, September 23.

Wolfe, David. 2001. *Tales from the Underground: A Natural History of Subterranean Life*. New York: Perseus.

Wolf-Meyer, Matthew. 2017. "Our Master's Voice, the Practice of Melancholy, and Minor Sciences." *Cultural Anthropology* 30, no. 4: 670–91.

Zeiderman, Austin. 2013. "Living Dangerously: Biopolitics and Urban Citizenship in Bogotá, Colombia." *American Ethnologist* 40, no. 1: 71–87.

Zeiderman, Austin. 2016. *Endangered City: The Politics of Security and Risk in Bogotá*. Durham, NC: Duke University Press.

Zibechi, Raúl. 2007. *Dispersar el Poder: Los movimientos como poderes antiestatales*. Quito: AbyaYala.

campesinos: (*continued*)
colonialism and, 35, 51, 113; dispossession of, 28, 122, 129–30, 142–43; food autonomy and, 148–52; identity of, 17–18, 156, 161; La Hojarasca and, 17–20, 30–31, 138; land restitution and, 24, 30, 83–84; scientific knowledge and, 6, 33–37, 76; as selvacinos, 152–61, 173; social movements and, 24, 35, 107–8, 143–45, 154–55, 159; soil knowledge of, 59–61, 71–77, 129; state policy and, 17–18, 142–43; stereotypes of, 17, 20. *See also* selva

capitalism, 6–7, 20–21, 32–37, 52–53, 61–63, 78–85, 99–103, 134–36, 170–72; transition from, 150–52. *See also* extractivism; neoliberalism

Casid, Jill, 192n3

Center for Research and Popular Education (CINEP), 16–17

Chaves, Margarita, 184n7

Choy, Timothy, 8

Civil Society of Putumayo (SOCIVIL), 149–50

Clastres, Pierre, 147

coca: affect and, 144–46; extractivism and, 4, 81–82; food autonomy and, 22, 149–51; paramilitaries and, 2–4, 77–79, 81–83, 149; production of, 12, 18–24, 144–45; transition from, 13–24, 40, 104, 112, 120–25, 181; US-Colombia policy and, 2–4, 11–14, 29, 81–83, 104. *See also* aerial fumigation; Plan Colombia; war on drugs

Cohen, Benjamin, 52

Colebrook, Claire, 195n2

Colombia: antidrug policy in, 2–4, 11–14, 29, 73, 81–83, 104; biodiversity in, 70–71, 83; colonialism in, 26–30, 51; history of soil science in, 44–45, 65–66; land mines in, 68; multiculturalism in, 142–43; transitional justice in, 5–6; war in, 2, 22, 29

Colombian Agricultural and Livestock Institute, 106–7

colonialism: agriculture and, 4, 51, 72–74, 109; decoloniality and, 13, 32–37, 107, 113, 129–34, 154–56, 162, 174; decomposition and, 64, 122; history in Colombia of, 26–30, 51; science and, 6, 32–37, 51, 72–74

Committee for the Integration of the Colombian Macizo (CIMAFUNDECIMA), 107–8

composition and decomposition: hojarasca and, 3, 10, 16–18, 39–42, 120, 174; as life politics, 105–6, 114, 118, 125–36, 160, 171, 175; place and, 85, 88, 145, 156; selva and, 7–8, 27, 32, 90–92, 156; soil and, 64

Connolly, William, 65

CORPOAMAZONIA, 25, 29, 156–57

Corsín Jiménez, Alberto, 192n8

Cortés, Abdón, 38, 65–66, 99–102

criminalized ecology, 16, 80

decolonial enactments of asymmetry. *See* analytical symmetry

de la Cadena, Marisol, 37, 156

Deleuze, Gilles, 75, 92, 161, 165, 184n9

Delgado, Ana, 107

DeSilvey, Caitlin, 39

Despret, Vinciane, 31–32

development: conceptions of life and, 85, 151–52; crop substitution and, 13–14, 21–24, 89–90, 104; extractivism and, 24, 77–81, 85, 89, 122, 156, 171–73; indigenous communities and, 77; selva and, 16–17, 103; soil science and, 4–5, 45, 52, 63, 72, 77–79, 99–100; war on drugs and, 80–81

Diprose, Rosalyn, 63

dreams, 142, 162–68, 175

Duarte, Carlos, 142–43

Dumit, Joseph, 64

Duque, Iván, 82–83, 176–77

Environmental Clinic, 23, 162–65

extractivism: agriculture and, 4, 21, 38, 102, 134–36, 145; colonialism and, 28–33, 38, 64, 122; development and, 85, 114, 122; industrial forms of, 7–8, 24, 61–64, 150, 156; soil science and, 46–47, 61–64; transition from, 151; war on drugs and, 79, 84–85. *See also* capitalism; neoliberalism

FARC-EP. *See* Revolutionary Armed Forces of Colombia–People's Army (FARC–EP)
farmers. *See* campesinos
farm school. *See* La Hojarasca
Federici, Silvia, 63
Food and Agriculture Organization of the United Nations (FAO), 49–52, 59, 100, 170–72
Franco, Fernando, 72
Freire, Paulo, 33

Gambetti, Zeynep, 134
Garcia, Angela, 191n7
García Márquez, Gabriel, 175
Gibson-Graham, J. K., 148
glyphosate, 2–3, 11, 13, 15, 38, 77–86, 176, 180. *See also* aerial fumigation; war on drugs
González, María Camila, 60
Grubačić, Andrej, 193n5
Guattari, Félix, 75, 92, 161
Gutiérrez, Laura, 107

Hacking, Ian, 43
Hage, Ghassan, 175
Haraway, Donna, 39, 43, 68–69
Hartigan, John, 8
harvesting knowledge, 157–61
Hekman, Susan, 63
Hetherington, Kregg, 191n21
Hole, Francis D., 62–63
hojarasca (litter layers), 3, 7–8, 18, 39–40, 114, 120–36, 174–77
Hupy, Joe, 188n25

IGAC. *See* National Geographic Institute Agustín Codazzi (IGAC)
indigenous communities: alternative agriculture and, 105–8, 122, 142–44, 152–59; colonialism and, 35, 51; dispossession of, 3–4, 20–23, 24, 28, 34–35, 66, 143–44; self-determination of, 151–52; soil knowledge of, 59–61, 77
Ingold, Tim, 50, 67
Institute for Rural Development (INCODER), 87–88, 190n10

James, William, 39
Jenny, Hans, 46, 63
Jiménez, Alcides (Padre Alcides), 111–12, 122–23, 130

Kirksey, Eben, 186n21
Kohn, Eduardo, 8, 159

La Cuchara, 149–50
La Hojarasca (farm school), 10–11, 15–22, 30–31, 39, 71–72, 141–42, 175–76
land: agronomy and, 4–5, 24, 36, 57–64, 77–79, 94–102; collective forms of, 143, 148; colonialism and, 27–28; development of, 23, 85–87; dispossession and, 26–28, 60, 64, 85, 106–7, 122, 130; extractivism and, 114, 122; labor as basis of property in, 73–74, 88; ownership of, 24–25, 60–61, 66, 73–74, 80–81, 143–44, 148, 163, 176; paramilitaries and, 15, 29–30, 60–61, 82–85, 112–14, 143, 149; productivity of, 8, 24–28, 46, 60, 78, 94–102, 156; restitution of, 24, 30, 83–84; social movements around, 36, 75, 85, 143–44, 148, 159, 173–76; as vital space, 7, 142–44, 176; war on drugs and, 82–85, 112–14. *See also* agriculture; selva; soil; territory
Latour, Bruno, 59, 96
leaders, 160–62

learning: dreaming and, 165–68, 175; knowing vs., 33, 37; La Hojarasca and, 109, 114, 138; lecturaleza and, 91–93, 127–28; un- and relearning and, 21, 90–93, 155–59, 165, 171–74

lecturaleza, 91–93, 127–28

living knowledge, 31, 157–61, 174

Marx, Karl, 52–53

Massey, Doreen, 50, 185n18

Mbembe, Achille, 33

McKenzie, Marcia, 50

McLean, Stuart, 8

Meuret, Michel, 31–32

Mesa Regional de Organizaciones Sociales del Putumayo, Baja Bota Caucana y Cofanía Jardines de Sucumbíos, Nariño (MEROS), 23–24, 31, 131, 151, 156, 161

minor science, 75, 101, 173

Myers, Natasha, 8, 108, 170, 187n16

National Agrarian, Ethnic, and Popular Strike (2013), 3, 24, 112, 151, 185n12

National Code of Renewable Natural Resources (1974), 53

National Geographic Institute Agustín Codazzi (IGAC), 4–7, 38, 41–53, 57–60, 68–78, 93–104, 172–73. *See also* agroecology; agrology; soil science

National Political Economic and Social Council (CONPES), 53–54

National Soil Science Laboratory, 5, 38, 41, 44–45, 61, 96–98

neoliberalism, 14, 22–23, 36, 85, 107, 134. *See also* capitalism; extractivism

Network of Guardians of the Seeds of Life, 23, 105–9

O'Hearn, Denis, 193n5

oil, 27–28, 62, 80–87, 140, 150–52, 160–63

Ojeda, Diana, 60

ojos para ella (eyes for her), 7, 21, 85–91, 174

Padre Alcides (Alcides Jiménez), 111–12, 122–23, 130

Papadopoulos, Dimitris, 64, 75, 175

paramilitaries: FARC-EP vs., 15, 29–30, 112, 117; land appropriation by, 60–61; public violence of, 106, 130; repression by, 23, 82–83, 122–23, 127, 154, 177. *See also* United Self-Defense Forces of Colombia (AUC)

Parreñas, Juno, 183n5

Pérez-Bustos, Tania, 191n24

Pignarre, Philippe, 32, 90, 176–77

Plan Colombia, 2, 11–14, 29

poetics of soil health, 7–9, 44, 61–65, 176

Povinelli, Elizabeth, 43, 131, 134, 186n1

Primavesi, Ana, 140–42

Proyecto Radargramétrico del Amazonas (PRORADAM), 71–72, 76–77

Puig de la Bellacasa, María, 54–56

pulsations of cosustainability, 113, 132

Putumayo: agroforestry in, 147–51; coca production in, 2–3, 12, 18–19, 29; colonialism and, 26–30; demographics of, 158, 185n15; development projects in, 14–24; FARC-EP in, 15, 18, 29–30; flooding in, 42; indigenous communities in, 184n7; infrastructure in, 16, 28; land ownership in, 24–25; paramilitaries in, 15, 29–30; soil quality in, 68; soil studies in, 74–78, 100–102; war on drugs in, 2–3, 80–83, 104

Raffles, Hugh, 70–71, 185n18, 192n9

Ramírez, María Clemencia, 28

Rancière, Jacques, 193n11

Revolutionary Armed Forces of Colombia–People's Army (FARC-EP), 2, 15, 123, 163, 173; demobilization of, 80–83; Padre Alcides and, 111–18; peace accords with, 5, 30, 61; Putumayo and, 15, 18, 29–30. *See also* United Self-Defense Forces of Colombia (AUC)

Ritz, Karl, 66

robust fragility, 40

Rodríguez-Giralt, Israel, 107

Rosero, Carlos, 190n12

rural communities. *See* campesinos

Sabsay, Leticia, 134

Sagan, Dorion, 170

Salgado Ruiz, Henry, 22, 193n5

sancocho (stew), 158–59

science: boundaries of, 6, 37, 74–75, 157; capitalism and, 6; colonialism and, 6, 32–37, 51; democratization of, 34–37; epistemology of, 37, 43, 156–58; minor forms of, 75, 101, 173. *See also* soil science

Scott, James, 193n10

seeds, 21–23, 91–94, 101, 105–15, 120–38, 144–49, 162, 174

selva: as alternative agro-life processes, 7–8, 21–23, 31–32, 36, 39, 103, 113–16, 122, 142, 154–55, 161–68, 174; apprenticeship and, 32, 90–91, 156, 161, 173; colonialism and, 27–28; composition and decomposition and, 8, 31–32, 90–92, 156; cultivation and, 21, 31–32, 36, 40–41, 88–90, 102–3, 120, 171; dreams and, 162–66; extractivism vs., 21; La Hojarasca and, 16–18, 39; regeneration and, 7, 31–32, 103, 113–16, 132, 156; social movements and, 122; soil and, 71–73, 99, 103, 133–36, 171, 176. *See also* agriculture; learning; ojos para ella (eyes for her)

selvacino, 152–61, 173

Serje, Margarita, 153

Serres, Michel, 16–17, 42, 174

Shiva, Vandana, 110

Simpson, Leanne, 175

soil: care of, 74–75; criminalization of, 77–80, 146; decolonialization of, 174; degradation of, 48–50, 60, 64, 133, 170–71; generosity of, 63–65; health vs. quality of, 61, 63–65, 172; as laborer, 51–53, 63; life and death of, 134; litter layer of, 3, 71, 171; ontologies of, 4–5, 42–43, 46–51, 55, 76, 99, 169–72; peace and, 172–73; poetics of the politics of, 7–9, 44, 61–65, 176; as public concern, 51–52; racialized views of, 72–73; regeneration of, 125–29, 133–34; relationality of, 53–56, 62–66, 99–104; selva and, 71–73, 99, 103, 133–36, 171, 176. *See also* agriculture; National Geographic Institute Agustín Codazzi (IGAC); soil science; Year of Soils in Colombia

soil science: agriculture and, 21–22, 32–34, 45–46, 52, 56, 70, 75–78, 99–102, 135, 170; Cortés and, 65–66; critique of, 74–75; genealogy of, 41–61; minor science of, 75, 101, 173; as research object, 4–7; selva and, 173; soil management and, 171–72; surveys and, 77–78; USDA taxonomy and, 38, 59, 71–72, 98–103. *See also* agroecology; agrology; National Geographic Institute Agustín Codazzi (IGAC)

Solnit, Rebecca, 168

Star, Susan Leigh, 54–56, 99–100

State Institute of Agriculture and Livestock Merchandising (IDEMA), 18–19

Stengers, Isabelle, 20, 32, 38, 90, 113, 135, 145, 171–72, 176–77

Stephenson, Niamh, 175

Stewart, Kathleen, 17

Stoetzer, Bettina, 192n3

Stoler, Ann, 64

Svampa, Maristella, 143

Tate, Winifred, 185n11

Taussig, Michael, 105

tenaces, 160–62

territory: administration of, 23–24, 77–80, 87, 153, 169–71; agronomy and, 4–5, 24, 36, 57–60, 64, 77–79, 94–102; campesino visions of, 70, 75, 93, 132, 155–57; colonialism and, 27–28; conflicts over, 5–9, 28, 67–68, 79, 142–44, 172, 176; degradation of, 40; dispossession and, 26–28, 60, 64, 85, 106–7, 122, 130, 136; emancipation of, 32, 113, 173; frontier of, 22, 129–30; modern relations of, 4, 25; paramilitaries and, 15, 29–30, 60–61, 82–85, 106–7, 112–14, 143, 149; peace and, 79; rights to, 24, 143–44; as vital space, 7, 142–44, 176; war on drugs and, 82–85, 112–14. *See also* agriculture; land; selva; soil
Tilley, Helen, 35
time, 42, 174–75
Todd, Zoe, 194n11
Tsianos, Vassilis, 175
Tsing, Anna, 41
Tuck, Eve, 50

United Self-Defense Forces of Colombia (AUC), 15, 29, 123
Uribe, Álvaro, 29
USAID, 2, 11–15, 17, 80, 104, 113–15

US-Colombia antidrug policy. *See* war on drugs
US-Colombia free trade agreement, 24, 52, 85, 101, 113

Valparaíso, 106–11
Vergunst, Jo Lee, 67
vital frequency, 140–41, 174–75
vital space, 7, 120, 137–47, 176
Viveiros de Castro, Eduardo, 76, 151
Voyles, Traci Brynne, 189n3

war on drugs: coca and, 2, 22–23, 29–30; criminality discourse of, 176; environmental destruction from, 2–4, 11–14, 22–23, 104, 130, 135; extractivism and, 2–4, 77–83, 122, 150; La Hojarasca and, 17; military interventions in, 2–4, 29, 74, 77, 82; soil science and, 73–74. *See also* aerial fumigation; Plan Colombia
WinklerPrins, Antoinette, 70–71
Witness for Peace (Acción Permanente por la Paz), 2, 11
Wolfe, David, 43
Wolf-Meyer, Matthew, 75

Year of Soils in Colombia, 4–7, 43–44, 47–51, 54–55, 60, 71, 170–72